T0295735

Parallel Programming for Modern High Performance Computing Systems

Parallel Programming for Modern High Performance Computing Systems

Paweł Czarnul
Gdańsk University of Technology, Poland

CRC Press is an imprint of the
Taylor & Francis Group, an **informa** business
A CHAPMAN & HALL BOOK

CRC Press
Taylor & Francis Group
6000 Broken Sound Parkway NW, Suite 300
Boca Raton, FL 33487-2742

Printed on acid-free paper
Version Date: 20171127

International Standard Book Number-13: 978-1-1383-0595-3 (Hardback)

Visit the Taylor & Francis Web site at
http://www.taylorandfrancis.com

and the CRC Press Web site at
http://www.crcpress.com

To my daughter Ala

Contents

List of figures

List of tables

List of listings

Preface

Parallel computing systems have recently become more and more accessible to a wide range of users. Not only programmers in high performance computing centers but also a typical consumer can now benefit from high performance computing devices installed even in desktop computers. The vast majority of new computers sold today feature multicore CPUs and GPUs which can be used for running parallel programs. Such usage of GPUs is often referred to as GPGPU (General Purpose Computations on Graphics Processing Units). Among devices announced by manufacturers are, for instance, a 7th generation Intel® Core™ i7-7920HQ CPU that features 4 cores with HyperThreading for 8 logical processors clocked at 3.1GHz (up to 4.1GHz in turbo mode) and a TDP (Thermal Design Power) of 45W. AMD Ryzen™ 7 1800X features 8 cores for 16 logical processors clocked at 3.6 GHz (4 GHz in turbo mode) and a TDP of 95W. A high end desktop Intel Core i9-7900X CPU features 10 cores with HyperThreading for 20 logical processors clocked at 3.3GHz (up to 4.3GHz in turbo mode) and a TDP of 140W. NVIDIA® Titan X, based on the NVIDIA® Pascal™ architecture, features 3584 CUDA® cores and 12GB of memory at the base clock of 1417MHz (1531MHz in boost) with a power requirement of 250W. Workstations or servers can use CPUs such as Intel® Xeon® Scalable processors such as Intel Xeon Platinum 8180 processor that features 28 cores and 56 logical processors clocked at 2.5 GHz (up to 3.8GHz in turbo mode) and a power requirement of 205W or Intel Xeon E5-4669v4 with 22 cores and 44 logical processors clocked at 2.2 GHz (3 GHz in turbo mode) and a TDP of 135W. AMD Opteron™ 6386 SE features 16 cores clocked at 2.8 GHz (3.5 GHz in turbo mode) with a TDP of 140 W. High performance oriented GPUs include NVIDIA® Tesla® P100, based on the Pascal architecture, with 3584 CUDA cores at the base clock of 1480MHz in boost and with a power requirement of 250W as well as NVIDIA Tesla V100 with 16 GB HBM2 memory, 5120 CUDA cores clocked at 1455MHz in boost and with a power requirement of 300W. AMD FirePro™ W9100 features 2816 Stream Processors, 32GB or 16GB GDDR5 GPU memory with a power requirement of 275W. High performance oriented machines can use coprocessors such as Intel® Xeon Phi™ x100 7120A with 61 cores clocked at 1.238GHz and a TDP of 300W or e.g. Intel Xeon Phi x200 7250 processors with 68 cores clocked at 1.4GHz with a TDP of 215W. As it was the case in the past and is still the case today, computer nodes can be interconnected together within high performance computing clusters for even greater compute performance at the cost of larger power consumption. At the time of this writing, the most powerful cluster on the TOP500 list is Sunway TaihuLight with over 10 million cores and over 93 PFlop/s Linpack performance. The cluster takes around 15.37 MW of power.

Hardware requires software, in particular compilers and libraries, to allow development, building and subsequent execution of parallel programs. There is a variety of existing parallel programming APIs which makes it difficult to become acquainted with all programming options and how key elements of these APIs can be used and combined.

In response to this the book features:

1. Description of state-of-the-art computing devices and systems available today such as multicore and manycore CPUs, accelerators such as GPUs, coprocessors such as Intel Xeon Phi, clusters.

2. Approaches to parallelization using important programming paradigms such as master-slave, geometric Single Program Multiple Data (SPMD) and divide-and-conquer.

3. Description of key, practical and useful elements of the most popular and important APIs for programming parallel HPC systems today: MPI, OpenMP®, CUDA®, OpenCL™, OpenACC®.

4. Demonstration, through selected code listings that can be compiled and run, of how the aforementioned APIs can be used to implement important programming paradigms such as: master-slave, geometric SPMD and divide-and-conquer.

5. Demonstration of hybrid codes integrating selected APIs for potentially multi-level parallelization, for example using: MPI+OpenMP, OpenMP+CUDA, MPI+CUDA.

6. Demonstration of how to use modern elements of the APIs e.g. CUDAs dynamic parallelism, unified memory, MPIs dynamic process creation, parallel I/O, one-sided API, OpenMPs offloading, tasking etc.

7. Selected optimization techniques including, for instance: overlapping communication and computations implemented using various APIs e.g. MPI, CUDA, optimization of data layout, data granularity, synchronization, process/thread affinity etc. Best practices are presented.

The target audience of this book are students, programmers, domain specialists who would like to become acquainted with:

1. popular and currently available computing devices and clusters systems,

2. typical paradigms used in parallel programs,

3. popular APIs for programming parallel applications,

4. code templates that can be used for implementation of paradigms such as: master-slave, geometric Single Program Multiple Data and divide-and-conquer,

5. optimization of parallel programs.

I would like to thank Prof. Henryk Krawczyk for encouraging me to continue and expand research in the field of parallel computing as well as my colleagues from Department of Computer Architecture, Faculty of Electronics, Telecommunications and Informatics, Gdańsk University of Technology for fruitful cooperation in many research and commercial projects, stating challenges and motivating me to look for solutions to problems. Furthermore, I would like to express gratitude for corrections of this work by Dr. Mariusz Matuszek, Paweł Rościszewski and Adam Krzywaniak. Finally, it was a great pleasure to work with Randi Cohen, Senior Acquisitions Editor, Computer Science, Robin Lloyd-Starkes, Project Editor and Veronica Rodriguez, Editorial Assistant at Taylor & Francis Group.

Paweł Czarnul
Gdańsk, Poland

CHAPTER 1

Understanding the need for parallel computing

CONTENTS

1.1 INTRODUCTION

For the past few years, increase in performance of computer systems has been possible through several technological and architectural advancements such as:

1. Within each computer/node: increasing memory sizes, cache sizes, bandwidths, decreasing latencies that all contribute to higher performance.
2. Among nodes: increasing bandwidths and decreasing latencies of interconnects.
3. Increasing computing power of computing devices.

It can be seen, as discussed further in Section 2.6, that CPU clock frequencies have generally stabilized for the past few years and increasing computing power has been possible mainly through adding more and more computing cores to processors. This means that in order to make the most of available hardware, an application should efficiently use these cores with as little overhead or performance loss as possible. The latter comes from load imbalance, synchronization, communication overheads etc.

Nowadays, computing devices typically used for general purpose calculations, used as building blocks for high performance computing (HPC) systems, include:

- Multicore CPUs, both desktop e.g. a 7th generation Intel i7-7920HQ CPU that features 4 cores with HyperThreading for 8 logical processors clocked at 3.1GHz (up to 4.1GHz in turbo mode) and server type CPUs such as Intel Xeon E5-2680v4 that features 14 cores and 28 logical processors clocked at 2.4 GHz (up to 3.3GHz in turbo mode) or AMD Opteron 6386 SE that features 16 cores clocked at 2.8 GHz (3.5 GHz in turbo mode).
- Manycore CPUs e.g. Intel Xeon Phi x200 7290 that features 72 cores (288 threads) clocked at 1.5 GHz (1.7 GHz in boost).
- GPUs, both desktop e.g. NVIDIA® GeForce® GTX 1070, based on the Pascal architecture, with 1920 CUDA cores at base clock 1506 MHz (1683 MHz in boost), 8GB of memory or e.g. AMD R9 FURY X with 4096 Stream Processors at base clock up to 1050 MHz, 4GB of memory as well as compute oriented type devices such as NVIDIA Tesla K80 with 4992 CUDA cores and 24 GB of memory, NVIDIA Tesla P100 with 3584 CUDA cores and 16 GB of memory, AMD FirePro S9170 with 2816 Stream Processors and 32 GB of memory or AMD FirePro W9100 with 2816 Stream Processors and up to 32GB of memory.
- Manycore coprocessors such as Intel Xeon Phi x100 7120A with 61 cores at 1.238GHz and 16 GB of memory or Intel Xeon Phi x200 7240P with 68 cores at 1.3GHz (1.5 GHz in boost) and 16 GB of memory.

1.2 FROM PROBLEM TO PARALLEL SOLUTION – DEVELOPMENT STEPS

Typically, development of computational code involves several steps:

1. Formulation of a problem with definition of input data including data format, required operations, format of output results.
2. Algorithm design. An algorithm is a procedure that takes input data and produces output such that it matches the requirements within problem definition. It is usually possible to design several algorithms that achieve the same goal. An algorithm may be sequential or parallel. Usually a sequential algorithm can be parallelized i.e. made to run faster by splitting computations/data to run on several cores with necessary communication/synchronization such that correct output is produced.
3. Implementation of an algorithm. In this case, one or more Application Programming Interfaces (APIs) are used to code the algorithm. Typically an API is defined such that its implementation can run efficiently on a selected class of hardware.

4. Code optimization. This step includes application of optimization techniques into the code that may be related to specific hardware. Such techniques may include data reorganization, placement, overlapping communication and computations, better load balancing among computing nodes and cores. It should also be noted that for some APIs, execution of certain API functions may be optimized for the given hardware in a way transparent to the programmer.

Parallelization of a sequential code may be easy as is the case, for example, in so-called embarrassingly parallel problems which allow partitioning of computations/data into independently processed parts and only require relatively quick result integration. The process is more difficult when control flow is complex, the algorithm contains parts difficult to parallelize or when the ratio of computations to communication/synchronization is relatively small. It is generally easier to parallelize a simpler sequential program than a highly optimized sequential version.

Typically parallelization aims for minimization of application execution time. However, other factors such as power consumption and reliability have become increasingly more important, especially in the latest large scale systems. Performance, power consumption and performance to power consumption ratios of the most powerful TOP500 [4] clusters are shown in Figure 1.1.

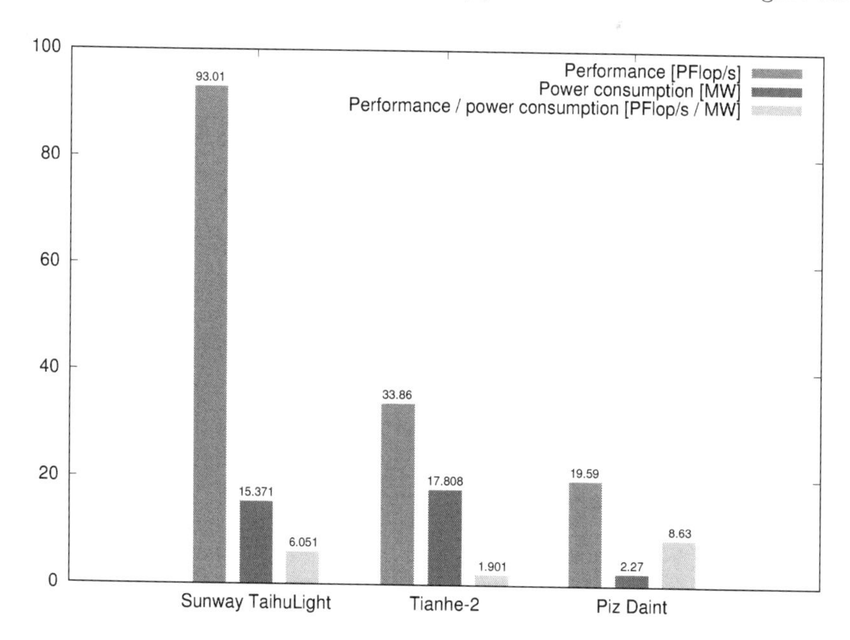

FIGURE 1.1 Top high performance computing systems according to the TOP500 list, June 2017, based on data from [4]

In spite of higher computational capabilities, power consumption of the first cluster on the list is lower than that of the second. Even better in terms of performance to power consumption ratio is Piz Daint that occupies the third spot. Such trends are shown in more detail in Section 2.6.

1.3 APPROACHES TO PARALLELIZATION

There are several approaches to programming parallel applications, including the following ones:

1. Programming using libraries and runtime systems that expose parallelism among threads, for instance: NVIDIA CUDA, OpenCL, OpenMP, Pthreads or among processes such as Message Passing Interface (MPI). In such cases, languages typically used for programming include C or Fortran. Such APIs may require management of computing devices such as GPUs (CUDA, OpenCL), handling communication of data between processes (MPI), between the host and GPU (unless new APIs such as uniform memory in CUDA are used).

2. Declarative in which existing sequential programs may be extended with directives that indicate which parts of the code might be executed in parallel on a computing device such as a multicore CPU or an accelerator such as GPU. Directives may instruct how parallelization is performed such as how iterations of a loop are distributed across threads executing in parallel, including how many iterations in a chunk are assigned to a thread (Section 4.2.2 shows how this can be done in OpenMP). It is also possible to mark code regions to ask the compiler to perform analysis and identification by itself (such as the `kernels` construct in OpenACC).

   It should be noted that APIs such as OpenMP and OpenACC, which allow directives, also contain libraries that include functions that can be called within code to manage computing devices and parallelism.

3. Programming using higher level frameworks where development of an application might be simplified in order to increase productivity, usually assuming a certain class of applications and/or a programming model. For instance, in the KernelHive [141] system an application is defined as a workflow graph in which vertices denote operations on data while directed edges correspond to data transfers. Vertices may be assigned functions such as data partitioners, mergers or computational kernels. Definition of the graph itself is done visually which makes development easy and fast. Input data can be placed on data servers. Computational kernels are defined in OpenCL for portability across computing devices and are the only snippets of code of the application a programmer is expected to provide. An execution engine, along with other modules, is responsible for selection of computing devices taking into account

optimization criteria such as minimization of application execution time [141] or minimization of application execution time with a bound on the total power consumption of selected services [55] or others.

Other frameworks or higher level abstractions for using computing devices include: rCUDA [63, 64] (it allows concurrent and remote access to GPUs through an API and makes it possible to use many GPUs in parallel), VirtualCL (VCL) Cluster Platform [19] (enables to run OpenCL applications in a cluster with potentially many computing devices) and Many GPUs Package (MGP) [18] (enables to launch OpenMP and OpenCL applications on clusters that include several GPUs), OMPICUDA [105] (allowing to run applications implemented with OpenMP/MPI on hybrid CPU and GPU enabled cluster systems), CUDA on Apache™ Hadoop® [3, 86] (two levels of parallelism are possible – using map-reduce at a higher and CUDA at a lower level, support for Java™ and C languages is available).

In terms of a programming level within a parallel application the following types of code can be distinguished:

- Single/uniform device type – one API is used to program a set of given devices such as MPI for a cluster that is composed of a number of nodes with CPUs or CUDA for programming an application running on one or more GPU cards. It should be noted that a particular API may allow for programming parallel code at various levels. For instance, CUDA allows management of computations among several GPUs from a host thread and parallel computations by many threads within each GPU.
- Hybrid – in this case one or more APIs are used to program computations and communication/synchronization among computing devices that may differ in computing power, power consumption, overhead for starting/termination of computations, performance in memory access, caching etc. For instance, OpenCL can be used for performing computations on multicore CPUs and GPUs installed within a workstation. On the other hand, APIs can be combined to allow programming such systems e.g. OpenMP can be used for running parallel threads on a host. Some threads may run computations in parallel on multicore CPU(s) and some may be responsible for management of computations on one or more GPUs. Such computations might be parallelized using CUDA.

A target system for parallelized code may be one of the following:

1. *Homogeneous* – when is contains many identical components at the same level i.e. a cluster may be composed out of a large number of identical computational nodes. Each node may include identical CPUs.
2. *Heterogeneous* – at one level a computational system may differ in computational power, power consumption etc. This makes parallelization

more difficult as it requires non-trivial partitioning of data/computations and synchronization among such systems. An example can be a computational server with multicore CPUs and various GPUs installed in it and further a collection of nodes with various computing devices.

1.4 SELECTED USE CASES WITH POPULAR APIS

Selected use cases demonstrating usage of particular APIs on specific computing devices are shown in Figure 1.2:

- 1 – A multithreaded application using OpenMP running on one or more CPUs within a node (desktop computer/workstation/server/cluster node).
- 2 – An application started and running on a host on a CPU that *offloads* a part of its code for execution on an Intel Xeon Phi coprocessor. Input data can be copied to the Xeon Phi and results can be copied back to the host.
- 3 – A multithreaded application using OpenMP running on cores of an Intel Xeon Phi coprocessor. The Intel Xeon Phi coprocessor runs Linux®.
- 4 – A parallel application that consists of typically two or more processes running on possibly several nodes. Processes may also run on various cores of CPUs installed within nodes. MPI also allows processes to spawn threads and use MPI functions to a degree supported by the given implementation (see Section 4.1.14).
- 5 – A parallel application that consists of typically two or more processes that, in this case, run on CPU cores of the host and Xeon Phi coprocessor cores in parallel. Note that performance of Xeon Phi cores and host CPU cores are typically different. This model is called *symmetric*.
- 6 – A parallel application that consists of typically two or more processes that, in this case, run on Xeon Phi coprocessor cores. This model is called *native*.
- 7 – An application started on a host CPU which runs some computations (coded within so-called kernels) on one or more GPU cards. Input data can be copied to a GPU(s) and results can be copied back to host RAM. The host code uses NVIDIA CUDA API.
- 8 – An application started and running on a host on a CPU that *offloads* a part of its code for execution on an accelerator or accelerators – in this case a GPU or GPUs. Input data can be copied to the GPU and results can be copied back to the host. OpenACC is used.

- 9 – An application started on a host CPU which runs some computations (coded within so-called kernels) on one or more GPU cards and/or host multicore CPUs. Input data can be copied to a GPU(s) and results can be copied back to host RAM. The host code uses OpenCL.

Figure 1.2 shows, at a high level, which types of systems are typical targets for parallel applications coded with a specific API considered in this book. The book also shows examples of combining APIs for hybrid parallel programming. As more and more HPC systems combine CPUs and accelerators, the presented templates and techniques are very likely to be applied to forthcoming systems as well.

1.5 OUTLINE OF THE BOOK

The outline of the book is as follows.

Chapter 2 discusses current state-of-the-art computing devices used in high performance computing systems including multicore CPUs, accelerators and coprocessors, available cluster systems. Furthermore, volunteer based computing systems are discussed that can be lower cost alternative approaches suitable for selected problems. In these cases, however, reliability of computations as well as privacy might be a concern. Finally, for completeness, a grid based approach is discussed as a way to integrate clusters into larger computing systems.

Chapter 3 first describes main concepts related to parallelization. These include data partitioning and granularity, communication, allocation of data, load balancing and how these elements may impact execution time of a parallel application. Furthermore, the chapter introduces important metrics such as speed-up and parallel efficiency that are typically measured in order to evaluate the quality of parallelization. The chapter presents main parallel processing paradigms, their concepts, control and data flow and potential performance issues and optimizations. These are abstracted from programming APIs and are described in general terms and are then followed by implementations in following chapters.

Chapter 4 introduces basic and important parts of selected popular APIs for programming parallel applications. For each API a sample application is presented. Specifically, the following APIs are presented:

1. Message Passing Interface (MPI) for parallel applications composed of processes that can exchange messages between each other. Depending on support from a particular implementation, threads can also be used within an application.

2. OpenMP for writing parallel multithreaded programs.

3. Pthreads for writing parallel multithreaded programs.

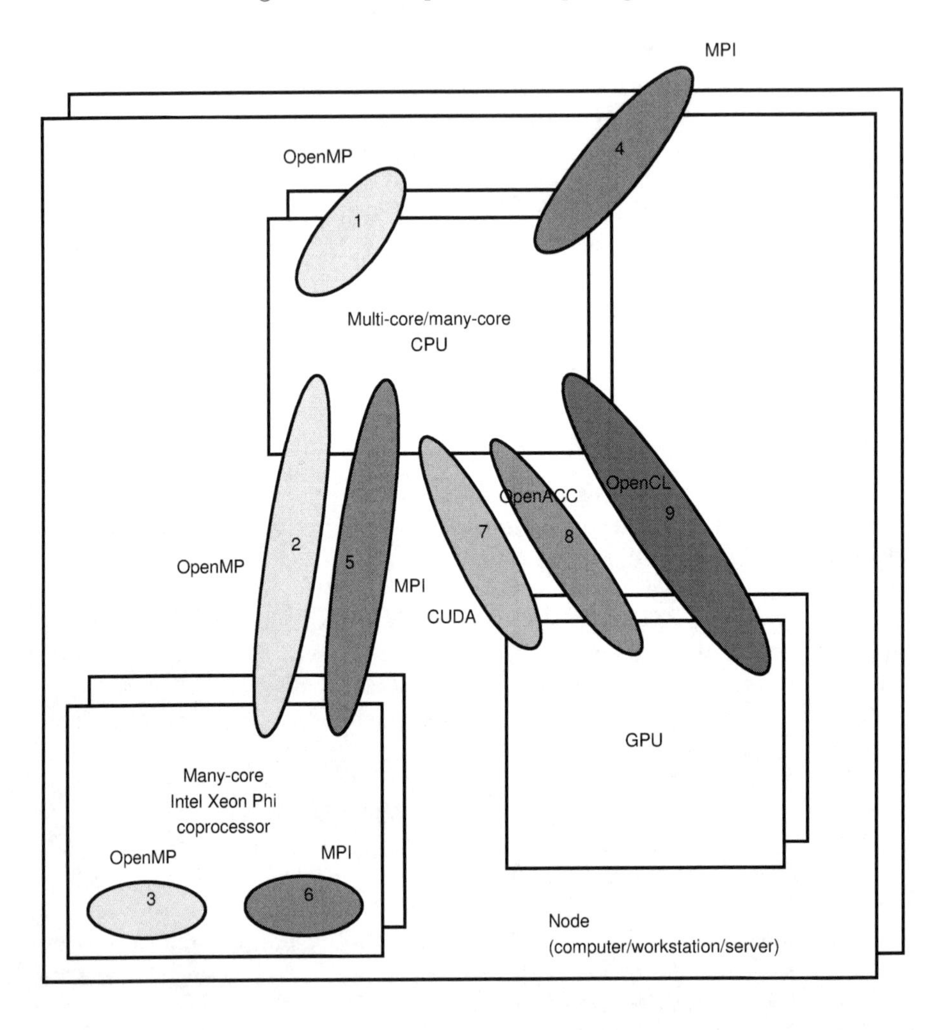

FIGURE 1.2 Typical usage of APIs for parallel programs on specific computing devices/architectures (other configurations might be possible depending on particular compilers/software)

4. CUDA for writing parallel multithreaded programs running on NVIDIA GPUs.

5. OpenCL for writing parallel multithreaded programs running on multicore CPUs and/or GPUs.

6. OpenACC for writing parallel multithreaded programs running on accelerators such as GPUs.

The chapter also describes examples of hybrid applications using selected combinations of APIs i.e. MPI+OpenMP, MPI+CUDA, MPI+Pthreads.

Chapter 5 presents implementations of master-slave, geometric SPMD and divide-and-conquer paradigms introduced in Chapter 3 using APIs or combinations of APIs presented in Chapter 4.

Finally, Chapter 6 describes selected techniques used to optimize parallel programs, including data prefetching, overlapping communication and computations, load balancing techniques, minimization of synchronization overheads. Data placement and thread affinity are discussed in the context of possible impact on performance. Furthermore, optimizations typical of GPUs, Xeon Phi, clusters and hybrid systems are presented.

CHAPTER 2

Overview of selected parallel and distributed systems for high performance computing

CONTENTS

This chapter describes selected state-of-the-art parallel systems as well as programming models supported by them. Next chapters describe specific APIs that implement these models and consequently allow use of these hardware architectures.

2.1 GENERIC TAXONOMY OF PARALLEL COMPUTING SYSTEMS

Nowadays, there are several types of systems that are used for high performance computing (HPC). The well known Flynn's taxonomy distinguishes systems by whether a single instruction stream (SI) or multiple instruction streams (MI) process either a single data stream (SD) or multiple data streams (MD) [67]. This results in SISD, SIMD, MISD and MIMD types of systems. In terms of memory-type systems, the following can be distinguished:

1. Shared memory systems – threads and processes may use shared memory for exchanging data and/or synchronization. There can be systems composed of many nodes as well as individual nodes that feature more and more cores in the following computing devices:
   (a) multicore CPUs such as Intel Xeon E5 or E7 or manycore CPUs such as Intel Xeon Phi x200,
   (b) accelerators and coprocessors including:
      - GPUs,
      - coprocessors such as Intel Xeon Phi x100 or coprocessor versions of Intel Xeon Phi x200.
2. Distributed memory systems – threads and processes of a parallel application generally do not have direct access to all memories of the whole system. Such systems include, in particular:
   (a) clusters – a cluster consists of many nodes interconnected with a fast network such as Infiniband™, nodes will typically include:
      - multicore CPU(s),
      - accelerator(s)/coprocessor(s),
   (b) volunteer computing – this type of system consists of a server or servers to which volunteers connect in order to download data sets, process the latter and send results back. This way, a large number of independent volunteers take part in large scale, usually socially oriented initiatives.

It should be noted that the term multicore usually refers to cores of traditional multicore CPUs that evolved from traditional CPUs by adding more and more cores. On the other hand, manycore is often referred to computing devices with tens or hundreds of cores designed with less powerful cores with an intent to run parallel applications.

The aforementioned systems can be described with the UML® notation as shown in Figure 2.1. The key difference between clusters and volunteer based systems in terms of hardware is geographical distribution of volunteers' nodes compared to centralized location of cluster nodes. Furthermore, clusters usually feature fast interconnects such as Infiniband compared to slow WANs in volunteer based systems. Additionally, volunteers are independent and codes run in an untrusted environment in contrast to a single cluster environment typically maintained by one administrator. In this context, a grid is typically referred to as a collection of clusters and servers, possibly in various administrative domains.

2.2 MULTICORE CPUS

Multicore CPUs have become standard not only in HPC systems but also in desktop systems. In fact, the number of cores per node is expected to grow to

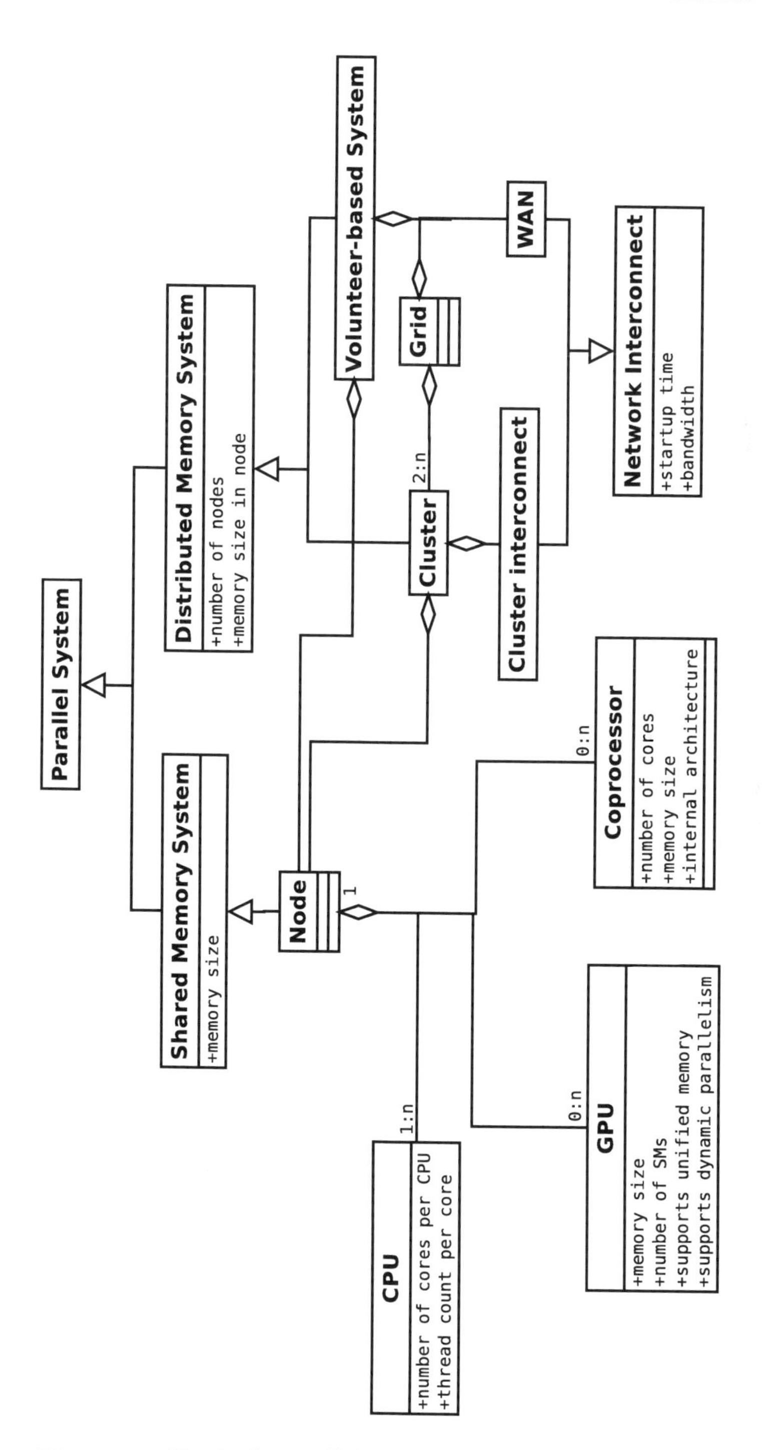

FIGURE 2.1 Typical parallel systems described by an UML diagram

numbers such as 128 [99] or more along with adoption of other technologies for storage such as NVRAM. Depending on a platform, a few CPUs per node could be incorporated into a node of an HPC system.

Examples of server type CPUs used today include Intel Xeon E7-8890 v3 that features 18 cores (36 threads with HyperThreading) clocked at 2.50 GHz (3.3 GHz in turbo) with a TDP of 165W or AMD Opteron 6386 SE that features 16 cores clocked at 2.8 GHz (3.5 GHz in turbo) with a TDP of 140W.

From the programming point of view, there are several APIs available for effective usage of multicore CPUs within a node, including in particular:

1. OpenMP [32, 129, Chapter 5] – allowing extension of a sequential application with directives that specify regions executed in parallel by several threads. A set of functions is available that allows threads to find out e.g. their own ids and the number of threads running in parallel in a given block of code.

2. OpenCL [17] – an application is represented in the form of a grid called NDRange which is composed of work groups each of which is composed of a number of work items. Work items in work groups and work groups in an NDRange can be logically aligned in 1, 2 or 3 dimensions.

3. Pthreads [115, 129, Chapter 4] – the standard allows creation of threads that execute designated functions. The Pthreads API allows synchronization using mutexes and condition variables that allow a thread to launch a blocking wait until a signal from another thread arrives (possibly after the other thread has provided new data for processing).

An advantage of such an environment is the relatively large memory size that can be installed within a node e.g. 512 GB or more. As of this writing, this is considerably larger than memory sizes available within accelerators such as GPUs described next.

2.3 GPUS

In today's landscape of computer systems, accelerators play an important role. Such computing devices offer considerable computational power for well structured computations compared to CPUs. Consequently, both CPUs and accelerators/coprocessors can be used together for parallel processing. As of now, the latter are generally plugged into motherboards into PCIe slots. Consequently, this communication link may become a bottleneck in case input data needs to be sent frequently from a host's RAM to an accelerator's memory for relatively short-lived computations.

The architectures of GPUs have been evolving from generation to generation. For instance, NVIDIA introduced architectures such as NVIDIA® Fermi™, NVIDIA® Kepler™, NVIDIA® Maxwell™, NVIDIA® Pascal™ and NVIDIA® Volta™.

NVIDIA CUDA application programming interface (API) presented in Section 4.4 is an abstraction allowing to implement programs that run on many architectures of NVIDIA cards. OpenCL presented in Section 4.5 allows implementation of programs also for GPUs by other manufacturers assuming proper support. It is important to realize how mapping of application elements is performed onto hardware and what limitations are considered by schedulers in order to maximize performance.

From the point of view of high performance computations, especially the following parameters of a GPU are important, using CUDA terminology:

- streaming multiprocessor – its architecture and the number of such elements in a GPU card,
- the number of FP32/FP64 CUDA cores per multiprocessor and the GPU card,
- maximum number of threads per multiprocessor,
- maximum number of threads per thread block,
- maximum number of thread blocks per multiprocessor,
- maximum number of registers per thread, per block, per multiprocessor,
- size of shared memory per multiprocessor,
- clock speed,
- memory bandwidth,
- memory size,
- cache size,
- support for atomic operations,
- TDP.

Article [65] compares specifications of GK180, GM200, GP100 and GV100 used in Kepler, Maxwell, Pascal and Volta generations respectively.

NVIDIA proposed some technologies to improve performance of its products, in particular:

1. NVIDIA GPUDirect™ – a technology that enables many GPUs or other devices such as network cards to access CUDA memory directly for reading/writing. A few technologies within GPUDirect are available such as [120]:
    - GPUDirect Shared Memory – allowing removal of copies of data in system memory when sending data out from a GPU's memory over a network.

- NVIDIA GPUDirect Peer-to-Peer – allowing optimized access to another GPU's memory within the same node.
- GPUDirect Support for RDMA – allowing remote access through a network to a GPU's memory on another node.

2. NVIDIA® NVLink™ – an interconnect technology that speeds up connectivity compared to PCIe 3.0. It allows direct connections between GPUs such as P100. Document [118] shows a configuration of 8 P100s connected with NVLink allowing applications to use memory of another GPU apart from the local resources. NVLink also allows connecting a GPU to a CPU such as POWER8® for increased bandwidth compared to PCIe 3.0.

3. Several new solutions adopted in NVIDIA Tesla V100 [65], in particular: mixed-precision FP16/FP32 Tensor Cores designed for deep learning, independent thread scheduling that allows interleaving of instructions from divergent paths in the code.

Other manufacturers offer other technologies for performance improvement, for example AMD DirectGMA for peer to peer PCI Express transfers without using host memory.

There are several cards available. For instance, NVIDIA GeForce GTX 1080 (Pascal architecture) features 2560 CUDA cores (clocked at 1607 MHz, 1733 MHz in boost), 8 GB GDDR5X with 320 GB/s bandwidth with a power requirement of 180W. AMD Radeon™ RX 580 features 36 Compute Units for a total of 2304 Stream Processors (clocked at 1257 MHz, 1340 MHz in boost), 8 GB GDDR5 with 256 GB/s bandwidth with a power requirement of 185W. NVIDIA Tesla K80 features 4992 CUDA cores, 24 GB GDDR5 memory with 480 GB/s bandwidth, with a power requirement of 300W. NVIDIA Tesla P100 (Pascal architecture) features 3584 CUDA cores, 16 or 12 GB CoWoS HBM2 memory with 720 or 540 GB/s bandwidth with a power requirement of 250W. NVIDIA V100 (Volta architecture) features 5120 CUDA cores, 16 GB HBM2 memory with 900 GB/s bandwidth with a power requirement of 300W. AMD FirePro W9100 features 44 Compute Units for a total of 2816 Stream Processors, 32GB or 16GB GDDR5 GPU memory with 320 GB/s bandwidth with a power requirement of 275W and OpenCL 2.0 support. NVIDIA DGX-1 is a deep learning system that features 8 Tesla GP100 GPUs for a total of 28672 CUDA cores, a Dual 20-core Intel Xeon E5-2698 v4 2.2 GHz, 512 GB 2133 MHz RAM, a dual 10 GbE and 4 IB EDR network interfaces with a power requirement of 3200W. NVIDIA DGX Station is another deep learning system that includes 4 Tesla V100 GPUs for a total of 20480 CUDA cores, 2560 Tensor cores, an Intel Xeon E5-2698 v4 2.2 GHz, 256 GB DDR4 RAM, a dual 10 GbE LAN with a power requirement of 1500W.

2.4 MANYCORE CPUS/COPROCESSORS

As outlined in Section 2.1, manycore computing devices refer to processors that typically feature a large number of cores, the latter relatively less powerful than the cores of high-end multicore CPUs. Manycore processors allow and require a high level of parallelism in an application for exploiting maximum performance. This section focuses on Intel Xeon Phi as a processor/coprocessor of this type.

The first generation Intel Xeon Phi x100 is a coprocessor that can be installed in a server in a PCI Express slot. It is an example of a manycore computing device that features several tens of physical cores, depending on the model. For instance, Intel Xeon Phi Coprocessor 7120A features 61 cores with the possibility to run 244 threads efficiently. The device has a TDP of 300 W. The global memory of the x100 series coprocessors can be as large as 16GBs. Each core has its L1 and L2 caches. The aforementioned Xeon Phi model has 30.5 MB L2. The internal architecture of the device features the following components:

1. cores with L1 (32KB data and 32 KB instructions) and L2 (512KB) caches,
2. memory controllers,
3. Tag Directories,
4. a ring interconnect to which the previous components are attached.

It is important to note that the x100 devices are installed in a computer in a PCIe slot which implies communication through PCIe which can be a bottleneck. The computational power of these coprocessors is around 1-1.2 TFlop/s depending on the model. Each core in these Xeon Phi devices is based on Intel Pentium core (with 64-bit support) and allows in-order execution only. Per core performance is lower than server multicore Xeon CPUs, also due to lower clock frequency for the Xeon Phi cores. From the point of view of power consumption, a Xeon Phi coprocessor can often be compared to two server type Xeon CPUs. Intel Xeon Phi supports efficient execution of four hardware threads per core. At least two threads per core should generally be used in order to utilize the computational potential of the hardware [135]. A core contains two pipelines which allow execution of two instructions per cycle (V- and U- pipelines).

Fundamentally, computational power of an Intel Xeon Phi stems from the following:

1. massive parallelization of the application itself, allowing a large number of threads to run in parallel on the many core hardware,
2. vectorization, allowing speed-up of up to 8 for double precision and up to 16 for single precision arithmetics,

3. use of instructions such as Fused Multiply Add (FMA) operations.

From this point of view, a programmer should make sure that the aforementioned features are utilized to a maximum extent in the application. Specific optimization points for this architecture are outlined in Section 6.9.2.

One of the advantages of programming for the Intel Xeon Phi platform is the possibility to use well established programming APIs including:

- OpenMP – as introduced before,
- MPI [6, 78, 77, 129, Chapter 3] – an application is composed of many processes running in parallel which exchange messages using, in particular, point-to-point or collective type operations. Processes can also launch threads that can invoke MPI functions in a way depending on support from a particular MPI implementation.

These are discussed in more detail in Chapter 4.

Each Intel Xeon Phi card runs Linux which also determines usage models/modes for nodes featuring Xeon Phi devices:

- *native* – Intel Xeon Phi is treated as a separate system onto which the programmer logs in and runs a parallel application on the Xeon Phi cores. This mode can be implemented with OpenMP or MPI.
- *offload* – an application is started on a CPU (there can be more than one CPU with multiple cores) and offloads some computations to an Xeon Phi coprocessor or more Intel Xeon Phi coprocessors if installed in the node.
- *symmetric* – an application runs both on cores of the host CPUs and cores of Intel Xeon Phi coprocessor(s). It should be kept in mind that efficiency of a CPU core and an Intel Xeon Phi core is different and this needs to be accounted for within the application that can be implemented with MPI in this case.

Additionally, other software APIs are available such as Intel Threading Building Blocks [91, 157], Intel® Cilk™ Plus [90, 157] or Intel hStreams for stream-based programming [92, 93, 97, Chapter 15].

Extending the Intel Xeon Phi family, Intel launched x200 series Intel Xeon Phi processors which are host bootable devices with high bandwidth integrated memory and out-of-order execution. As an example Intel Xeon Phi 7290 processor features 72 cores clocked at 1.5GHz and a TDP of 245W. Article [142] presents architecture of a processor in which 72 cores (288 logical processors) are laid in 36 tiles (each contains 2 cores and L2 cache) which are connected with a mesh. Apart from DDR4 memory, high bandwidth MCDRAM is also available. As discussed in Section 6.9.2, one of several cluster and one of several memory modes can be set. The former defines how cores are grouped into clusters. The latter defines how DRAM and MCDRAM are available to the application and how they can be used. Coprocessor versions of the x200 series were also developed.

2.5 CLUSTER SYSTEMS

Clusters have become the current incarnation of supercomputers – relatively easy to build and cost effective. From the programmer's point of view, parallelization is usually exploited at two levels:

1. Many cores of a computational device such as a multicore CPU, accelerator such as a GPU or a manycore computing device such as an Intel Xeon Phi coprocessor,
2. Many nodes of a cluster. Such nodes are typically interconnected with Gbit Ethernet and/or Infiniband, the latter offering very low latency of around 1-5us and bandwidth of e.g. almost 300 Gbit/s for EDR, 12x links.

As of this writing, according to the TOP500 [4] list, the top cluster system is Sunway TaihuLight using Sunway SW26010 260C processors clocked at 1.45GHz with Sunway interconnect. The cluster features 10649600 cores, offers 93.01 PFlop/s performance with power consumption of 15.37 MW. While this cluster uses manycore processors with 260 cores each, clusters that occupy the next three spots on the aforementioned list use combinations of multicore CPUs and either coprocessors or GPUs. These show current trends in HPC system architectures. Performance, power consumption and performance to power consumption ratios for the top three TOP500 clusters are shown in Figure 1.1.

In terms of programming APIs for clusters, typically implementations of the Message Passing Interface (MPI) standard [6] are used. MPI allows running a parallel application on both distributed memory (cluster nodes have their own RAM memories) and shared memory systems. Typically, an MPI application would consist of several processes that can exchange data by sending messages, either using point-to-point (Section 4.1.5) or collective communication calls (Section 4.1.6). Processes may launch threads that, depending on support from the MPI implementation, can also make MPI calls (Section 4.1.14). MPI can be used for communication among nodes and can be combined with other APIs for parallelization within a node, for example:

1. using multiple cores of node CPUs by launching threads with OpenMP [32, 127] or Pthreads [115],
2. launching computations on GPUs with NVIDIA CUDA [145, 122] or OpenCL [17],
3. launching computations on coprocessors such as Intel Xeon Phi with OpenMP.

2.6 GROWTH OF HIGH PERFORMANCE COMPUTING SYSTEMS AND RELEVANT METRICS

The recent history of parallel computing has seen a constant growth in performance of high performance computing systems as shown in the TOP500 [4] list. The latter is issued twice a year, in June and in November. Table 2.1 presents how the performance of the top cluster on the list changed every three years. The latter period is often considered for upgrade or replacement of typical computer systems.

It is important to note that nowadays this growing performance comes mainly from increasing the number of cores in such HPC systems as shown in Table 2.2. This is especially visible in the increasing number of nodes with not only multicore CPUs but also many core devices/accelerators.

Interestingly, clock frequencies of CPUs used in these modern HPC clusters have not changed considerably and actually decreased over the last years as indicated in Table 2.3. This is a serious challenge for programmers because they need to expose more parallelism within parallel implementations of algorithms and minimize communication and synchronization such that high speed-ups are obtained on a large number of cores.

TABLE 2.1 Performance of the first cluster on the TOP500 list over time, based on data from [4]

Date	Performance [TFlop/s]	Change in 3 years $\left[\frac{\text{current}}{\text{previous}}\right]$
06/2017	93014.6	2.75
06/2014	33862.7	4.15
06/2011	8162	7.96
06/2008	1026	← baseline

TABLE 2.2 The number of cores of the first cluster on the TOP500 list over time, based on data from [4]

Date	Number of cores	Change in 3 years $\left[\frac{\text{current}}{\text{previous}}\right]$
06/2017	10649600	3.41
06/2014	3120000	5.69
06/2011	548352	4.48
06/2008	122400	← baseline

Historically, taking into account the history of the TOP500 list, the most powerful cluster from this list would drop to the 500-th place on that list after

TABLE 2.3 CPU clock frequency of the first cluster on the TOP500 list over time, based on data from [4]

Date	Clock frequency [MHz]	Change in 3 years $\left[\frac{\text{current}}{\text{previous}}\right]$
06/2017	1450	0.66
06/2014	2200	1.1
06/2011	2000	0.63
06/2008	3200	← baseline

TABLE 2.4 Performance to power consumption ratio of the first cluster on the TOP500 list over time, based on data from [4]

Date	Performance/power consumption [MFlop/s / Watt]	Change in 3 years $\left[\frac{\text{current}}{\text{previous}}\right]$
06/2017	6051.3	3.18
06/2014	1901.54	2.31
06/2011	824.56	1.89
06/2008	437.43	← baseline

roughly 7-8 years. Furthermore, now, more than ever, energy consumption of such large systems and energy efficiency becomes a real concern. Specifically, it is aimed for that future clusters should not exceed 20MW of power consumption [61]. This is mainly due to high running costs of such systems. As shown in Table 2.4, the performance to power consumption ratio of the top cluster on the TOP500 has visibly increased.

Furthermore, increasing the size of a cluster in terms of the number of computational devices/the number of nodes brings the following matters into consideration:

1. Communication overhead and its impact on the speed-up of computations. For instance, communication start-up times are crucial for geometric Single Program Multiple Data (SPMD) parallel applications that simulate various phenomena such as medical simulations, weather forecasting etc. If a domain, in which a phenomenon is simulated, is partitioned into a very large number of regions each of which is assigned to a process/thread running on a distinct CPU or core, then the network startup time will be an important factor limiting the speed-up.

2. Reliability – given a large number of nodes probability of failure cannot be ignored [60]. This can be mitigated by checkpointing, described in Section 6.7.

3. Power consumption of computing devices. In general, today Intel Xeon E7 v4 server CPUs have power consumption up to around 165W while powerful GPUs such as NVIDIA Tesla K80 300W, NVIDIA Tesla P100 250W, NVIDIA Tesla V100 300W and Intel Xeon Phi x100 coprocessors up to 300W as well as Intel Xeon Phi x200 processors up to 275W. Consequently, from this point of view it is valid to compare performance of two server multicore CPUs against a GPU or a Xeon Phi as power consumption values are similar. Power consumption of large clusters will be associated with high running costs that cannot be neglected.

2.7 VOLUNTEER-BASED SYSTEMS

The main idea behind so-called volunteer computing is to engage independent, possibly geographically distributed, volunteers in a common computationally intensive task. Volunteer computing can be seen as an alternative to computing on high performance clusters for the class of problems in which computations and data can be divided into disjoint and independent fragments. There are, however, key differences that limit the type of computations that can be effectively processed in this manner or their performance:

1. A processing paradigm, in which volunteers only connect to a project server(s), download data packets for processing and upload results. No communication between volunteers is generally considered.
2. Geographical distribution of volunteers which results in:
    - considerable communication latencies between a volunteer and a project server as well as latency variation,
    - risk that a volunteer terminates processing at any time without even sending out a result – this causes the server to resend a task after a timeout,
    - risk of reporting a wrong result by a volunteer which effectively requires the server application to employ redundancy tactics mitigating the issue.

There are several frameworks and systems available that have incorporated the idea of volunteer based computing:

1. BOINC [13, 160] is one of the best known systems of this type. Volunteers, who would like do join one of several projects available, need to download and install client code which would then contact a project server to download input data for processing. Once processing has been finished, results are sent back to the server. The process is repeated allowing many independent volunteers to process data. It has been demonstrated that BOINC scales well as the BOINC distribution mechanisms can process ver 23 million tasks per day [14].

2. Comcute [16, 51] extended the basic idea implemented in BOINC with the following features:

   - In order to free the volunteer from the need for installation of software on the client side, by design computations are performed within a web browser. However, this potentially limits performance due to technologies that can be used for computations. The design of Comcute allows selection of potentially the best technology out of those supported on the client side. Technologies such as JavaScript® and Java applets were tested. Comcute still allows the running of dedicated clients, if needed.
   - A multi-level architecture that distinguishes the following layers (Figure 2.2):
     (a) Z layer – allows definition of data partitioners, data mergers, tasks with codes in various technologies that will be sent to volunteers for processing.
     (b) W layer – upper level task management in which a group of W servers can be selected that will manage execution of a given task collectively. This results in:
        i. Reliability – in case of W server failure, others can continue.
        ii. Better parallelization – particular W servers can be located in various geographical regions. Definition of a task includes: volunteer code, data partitioner and merger codes, input data as well arguments such as a desired level of redundancy. Then W servers partition input data and wait for the lower layer to prefetch data packets.
     (c) S layer – includes so-called distribution servers that are proxies between volunteers who join the system and W servers that host data packets. Several S servers can connect to a W server and many volunteers can communicate with an S server. S servers shall prefetch data packets from W servers and fetch new packets as volunteers request those.
     (d) I layer – volunteer layer: independent volunteers connect to publicly available Comcute's website [1] and upon joining computations are redirected to one of S servers from which their browsers fetch input data packets for processing and to which results are returned.

3. CrowdCL [111] is framework supporting volunteer computing and processing when visiting a web page. This open source solution allows using OpenCL, WebCL, JavaScript, CrowdCLient, KernelContext on the client in order to make use of available computing devices.

4. WeevilScout framework [37] features computing within a browser using Javascript.

5. Volunteer computing extended with processing using mobile devices. In [72] the authors distinguish task distribution point, task execution point, task distribution and execution point which can be assigned to clients. Using an example for prediction of a protein structure, execution times and energy consumption for such a distributed system are presented.

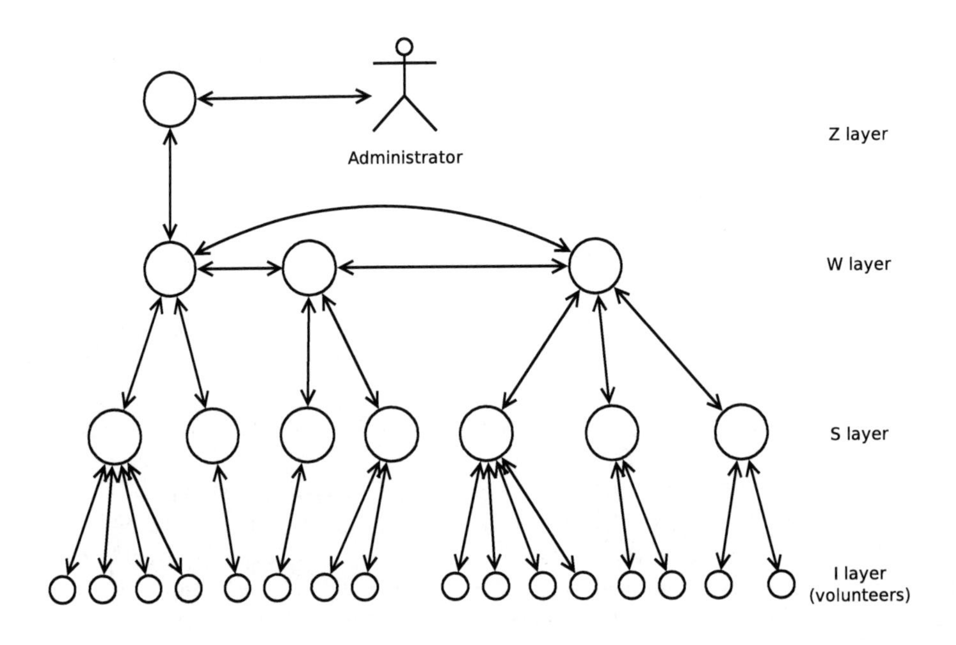

FIGURE 2.2 Architecture of Comcute

Cluster and volunteer based systems are clearly compute oriented but are really best suited for various application paradigms. Volunteer based systems are suitable for compute intensive applications in which independently processed data chunks can be easily distinguished, usually corresponding to either parts of a single solution or solutions for various input data sets. It should be noted that in volunteer computing:

1. A volunteer can intentionally terminate processing at any time. This can decrease computational power of a system.

2. Communication between a server and a volunteer can take considerable time. Unless data prefetching is implemented, idle times may occur.

3. Failures of nodes may occur or a volunteer may cheat. Again, this would decrease computational power of the whole system as the former would entail timeouts for detection of such a situation and both would require resending the same data packets to other volunteers.

Nevertheless, it is possible to compare computational power of cluster and volunteer based systems. At the time of this writing, according to the BOINC website [160], reported computational power (24-hour average) is approximately 15.832 PFlop/s thanks to over 186 thousand volunteers and over 1018 thousand computers compared to over 93 PFlop/s of the Sunway TaihuLight system.

Furthermore, it is interesting to compare not only performance but also the performance/power consumption ratio of both systems. In paper [54], based on statistics from the BOINC system for selected projects such as Atlas@Home, Asteroids@Home and also based on cross-project statistics, the following were computed:

1. expected CPU power consumption for a volunteer computer based on manufacturers' TDP values and CPU models active in particular projects,

2. expected CPU computational power based on PassMark® Software CPU Mark benchmarks [131] for CPU models active in particular projects.

Consequently, computational efficiency i.e. ratio of the expected computational power to expected power consumption was computed. For volunteer based systems, this resulted in the range between 45.3 to 76.3 CPU Marks/W compared to 145.7 for cluster Tryton located at Academic Computer Centre in Gdańsk, Poland (based on Intel Xeon E5-2670v3 CPUs, using InfiniBand FDR and running Linux) and 181.2 for cluster Cray® XC30™ located in the United States (based on Intel Xeon E5-2697v2 2.7GHz CPUs, using an Aries™ Interconnect and running the Cray Linux Environment).

Additionally, it can be estimated how much computational power is available in a volunteer system when certain parameters of such a system are assumed related to failures of processing input data packets. Specifically, in [53], it is assumed that an application requires processing of a certain number of data packets. For a single data packet, an estimated amount of work required to process a single data unit is computed taking into account: probability distribution of computational power of volunteers computers, probabilities of success and failure of processing if a volunteer is not able to complete processing within an imposed deadline.

2.8 GRID SYSTEMS

Grid computing technologies are to enable controlled resource sharing [69]. Clusters can be combined into grid [71, 70, 106, 152] systems, even if clusters are located in distinct administrative domains. The latter may correspond to various operating systems, file systems, queuing systems etc. Usually, grid middlewares, as shown in Figure 2.3, are used in order to hide these differences below a uniform layer that allows submission of jobs, file management and support proper security mechanisms. In essence, distributed HPC systems can

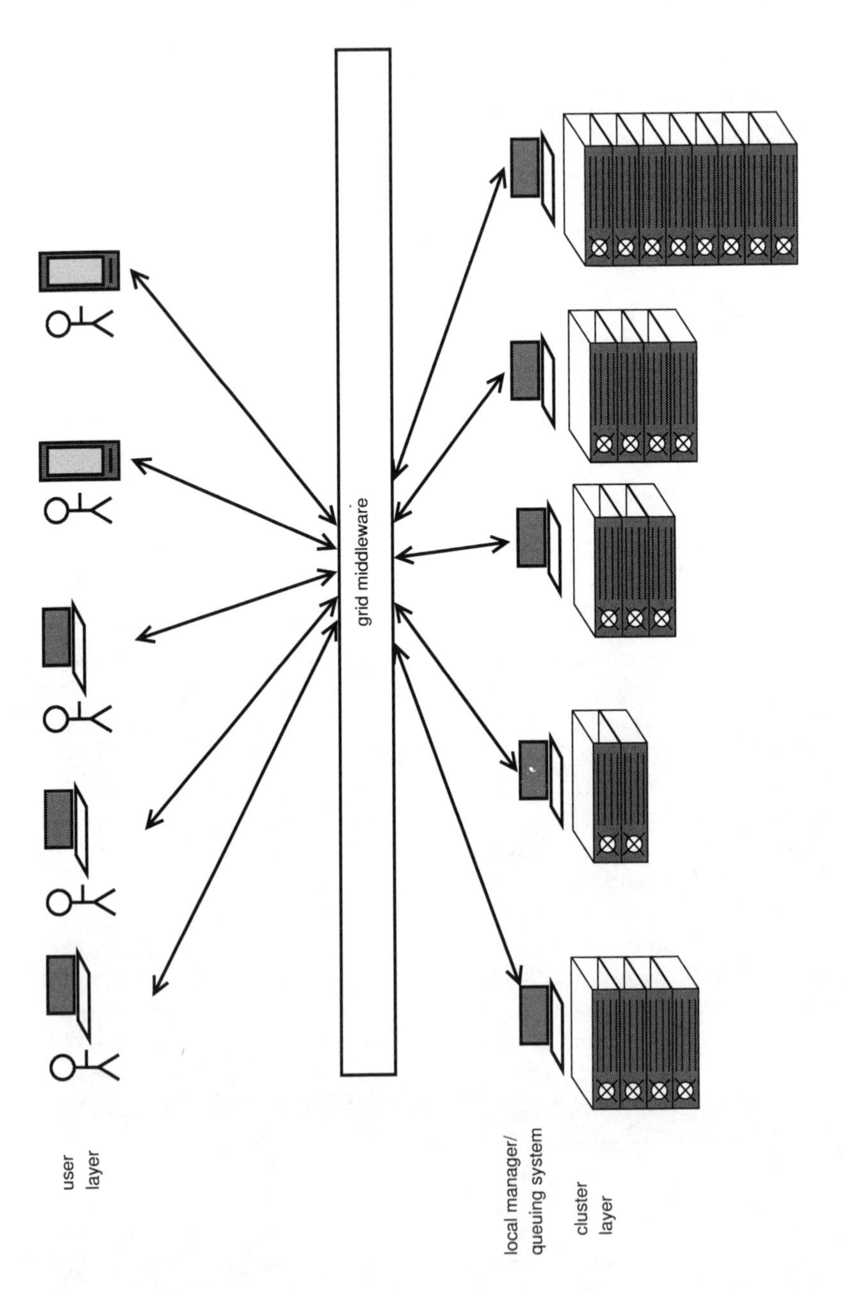

FIGURE 2.3 Grid computing

be integrated into one system so that a client can use the resources potentially transparently. Examples of grid middlewares include Globus® Toolkit [161], UNICORE [159] or Gridbus [30].

From a user's point of view, the following services are usually expected:

1. Job management including submission of jobs, querying the current status of a job, terminating a job. From the client point of view, resources in the grid may be transparent e.g. the system may find an appropriate server or a cluster that is compatible with the provided binary etc. In Globus Toolkit 6.0 GRAM [162] plays this role.
2. File management allowing copying of input/output data/results, in Globus Toolkit 6.0 possible through GridFTP [163].
3. Measures allowing secure communication with the grid e.g. using certificates. In Globus Toolkit authentication is performed using X.509 certificates through Grid Security Infrastructure (GSI) [164].

On the other end, various so-called virtual organizations may expose their resources to the grid. These resources, such as clusters, may use various technologies and software for:

- user authentication and authorization,
- queuing systems,
- file systems for user home directories and scratch spaces.

These differences should not absorb a grid client. Because of that, grid middleware allows to hide these differences behind a uniform interface that exposes desired services to users. Essentially, such services can be called directly or service invocations can be incorporated into higher level systems or modules, often with a graphical interface, allowing management of specific applications to be run on the grid.

An example of a grid system with a WWW and Web Service interface is BeesyCluster [40, 41, 48, 43] deployed as an access portal to high performance computing clusters of Academic Computer Center in Gdańsk, Poland. It allows users to set up an account that might be connected to an ACC system account but also accounts on other clusters and perform tasks such as editing programs, compilation, running either sequential or parallel jobs from command line or submitting jobs via queuing systems such as PBS, LoadLeveler® or others. Furthermore, BeesyCluster allows users to couple parallel applications from multiple nodes and clusters in a workflow application [44].

CHAPTER 3

Typical paradigms for parallel applications

CONTENTS

This chapter outlines important application paradigms, known in the context of parallel programming [29, Chapter 1], that are typical of many problems from various domains. Conversely, many specific problems can be addressed using templates corresponding to these paradigms. Specific parallelization techniques developed for these templates can then be used for corresponding applications. The techniques presented in this chapter abstract from particular programming APIs. On the other hand, Sections 5.1, 5.2 and 5.3 contain implementations of these techniques using modern parallel programming APIs. This chapter includes description and discussion on the following:

1. Basic terms used in parallelization of algorithms.
2. Parallelization approaches to typical programming paradigms including:
   - Master-slave – a master partitions input data and manages distribution of input data packets among slaves and collection of results and the subsequent merging phase.
   - Geometric parallelism (SPMD) – applications that typically simulate phenomena in 1D, 2D or 3D domains. A domain is typically

partitioned into a large number of cells each of which is described by a set of variables and their values with a meaning valid in the context of the problem simulated. A simulation progresses in time steps in which values of each cell are updated based on previous values of its neighbors and possibly the previous value for this very cell.

- Pipelining – input data consists of a (large) set of data packets/chunks each of which is processed in a series of stages. Each of the stages is executed in parallel on various data packets that are shifted through the stages in successive steps.
- Divide-and-conquer – an initial problem (and possibly data associated with it) is divided into a number of subproblems that can be solved in parallel. Each of the subproblems is either recursively divided into smaller subproblems or solved using a certain algorithm. Consequently, processing requires partitioning of problems and merging results until a final solution is obtained.

3.1 ASPECTS OF PARALLELIZATION

Typical goals of parallel programming are as follows:

1. Performance – in order to shorten execution time of an application compared to running a solution to the same problem in a sequential implementation.
2. Possibility to run – in some cases, resources such as memory or disk space of a single computer may not be enough in order to run an application. In a parallel system the total combined resources will often allow a parallel solution to run successfully.
3. Reliability – either several instances of the same code or instances of various parallel solutions to a problem might be run. This allows verification of either computations or algorithms/computational methods in a parallel environment.

There are several parallelization related concepts that show up during parallelization. These are discussed next.

3.1.1 Data partitioning and granularity

Data partitioning determines how input as well as intermediate data is divided among processes or threads working in parallel. This may be relatively easy if an application processes independent data chunks. However, if processing of certain data chunks depends on processing of others, i.e. explicit dependencies are involved, this may make the problem difficult to parallelize. Synchronization and communication between processes or threads involves overhead. Data partitioning can be:

1. static – if it is/can be done before processing starts,
2. dynamic – if new data chunks/packets are generated at runtime.

It should be noted that partitioning of input data in an algorithm can generate either:

1. A fixed/predefined number of data chunks/packets/subdomains (at each step of the algorithm where partitioning is involved) – which is determined by the algorithm itself. An example would be alpha beta search in a game tree in which the number of data packets/chunks would correspond to the number of positions generated from a given position – the number of legal moves.
2. A possibly variable number of data chunks/packets – in this particular case, the algorithm may adjust the number of data chunks/packets. For instance, searching for templates within a large text file: the input file can be partitioned into a certain number of chunks which are processed by processes/threads in parallel.

In the latter case, adjustment of granularity by setting the size of a packet typically results in a trade-off between:

- overhead for data packet/chunk management – typically a larger number of packets/chunks would result in additional time spent on synchronization and communication,
- imbalance – a smaller number of data packets/chunks would make it more difficult to balance load, especially if:
    1. data packets require various amounts of time to process,
    2. the number of data packets is similar to the number of processing cores and does not divide equally by the latter,
    3. the number of available processing cores exceeds the number of data packets.

Such trade-offs can be observed at various levels of parallelism i.e. between processor cores, cluster nodes and clusters. For instance, in [39] it is demonstrated how execution time of a parallel compute intensive workflow implementing adaptive quadrature integration for a given range depends on how input data (input range) was partitioned. For 16 cluster nodes, the input range was partitioned into various numbers of files with subranges. With processors of various speeds, a reasonably large number of files is needed for load balancing, but too large a number of files results in additional overhead.

A similar trade-off is shown in Section 6.2 for a parallel master-slave MPI application.

3.1.2 Communication

Naturally, in parallel and distributed systems data is typically distributed among processes or threads running on various cores and possibly nodes. An algorithm determines data and control flow as well as proper synchronization. Communication and synchronization involve overheads, depending on whether processes or threads run on a shared memory machine within a node (including multicore CPUs but also manycore systems such as Intel Xeon Phi), multiple cluster nodes, multiple clusters or hybrid environments including many levels of parallelism.

Modeling communication costs has been widely analyzed in the literature. Such models and coefficients would depend on the underlying hardware as well as software stacks. For instance, point-to-point communication time of message of size d can be modeled [146] as:

$$t_c(d) = t_s + \frac{d}{B} \tag{3.1}$$

where t_s denotes the startup time while B denotes the maximum possible bandwidth.

Paper [133] contains formulas describing communication times of various MPI calls as well as coefficients in the formulas (for a given MPI implementation and a cluster environment). Additionally, formulas for power consumption required by computations on a certain number of nodes with a given number of threads are presented along with dependency of power consumption per processor versus the number of active threads.

3.1.3 Data allocation

Data allocation defines how particular data packets/chunks are assigned to processes/threads for computations. Similarly to data partitioning, allocation can be:

1. Static – when data packets are assigned to processes/threads before execution of an algorithm starts. This can be the case e.g. in geometric SPMD applications in which each process/thread computes its own subdomain throughout the simulation loop [50].

2. Dynamic – when data packets are assigned to processes/threads at runtime.

Table 3.1 defines typical data partitioning/allocation combinations for the parallel processing paradigms discussed in this chapter. Specifically:

1. Master-slave – data is typically divided when an application starts or it is known it advance how input data is to be divided into data packets. Allocation is either static or dynamic (in which case slaves are assigned new data packets following completion of processing of previous data packets).

2. Geometric SPMD applications – an input domain is divided into subdomains either statically such as in FDTD methods [56, 169] or dynamically at runtime such as in MRTD methods [148, 169] when subdomains need to be redefined for load balancing. Typically though, allocation is static as it is beneficial to minimize data migration between e.g. cluster nodes.

3. Pipelining – data is initially divided into data packets/chunks which are then processed through pipeline stages – we assume that those stages are already assigned to particular processors or cluster nodes.

4. Divide-and-conquer – data is usually partitioned at runtime as the algorithm unfolds subproblems (such as in quick sort or alpha beta search algorithms) and needs to be allocated dynamically for efficient load balancing [47], at least for irregular applications [38].

TABLE 3.1 Typical data partitioning and allocation procedures in parallel processing paradigms

	data partitioning	**data allocation**
master-slave/task farming	static	static or dynamic
geometric SPMD applications	static or dynamic	static
pipelining	static	static
divide-and-conquer	dynamic	dynamic

3.1.4 Load balancing

Load balancing is the process of migrating data/work across available computing devices such as CPUs or accelerators so that each of these spends (approximately) the same amount of time to process the data. It should be noted that this is closely related to data partitioning and data granularity, especially in environments where there are multiple computing devices of varying speeds. Data granularity may determine how perfect load balancing can be achieved and what differences in execution times across computing devices arise. Additionally, if there are dependencies between processes or threads, synchronization and/or data communication will be needed. Consequently, this influences total application execution time.

There are many load balancing algorithms and approaches, targeted for various types of systems, both parallel and distributed. For instance, works [168, 171] present strategies and algorithms for parallel systems, [132] techniques for grid while [116] for cloud systems. Load balancing can be:

- Static – usually performed once before main computations start.

- Dynamic – performed periodically in response to changing load in order to balance data/computations among computing devices. In this case, every time load balancing is performed, its time will need to be added to the overall application execution time.

3.1.5 HPC related metrics

There are several metrics that define how well an application performs in a parallel environment.

Speed-up assesses how well the application scales when adding new computing devices to the environment. More specifically, assuming a number of identical computing devices, speed-up $S(N)$ can be defined as:

$$S(N) = \frac{t(1)}{t(N)} \tag{3.2}$$

where:

- $t(1)$ – execution time of the sequential application running on a system with 1 computing device,
- $t(N)$ – execution time of the application on a system with N computing devices.

Theoretically, $S(N)$ could reach the maximum value of N if we disregard any overheads such as communication, synchronization and if we assume perfect load balancing which is rarely the case in practical applications.

In some rare cases, superlinear speed-up may be observed in practice. However, this is usually the case when:

1. The environments are not equal when running the application on 1 computing device and on N computing devices. For instance, the environment with 1 computing device may have limited memory. For a memory hungry application this may result in the need for using a swap file and considerable slow down and consequently large execution time. On the other hand, a multi node environment may have considerably larger memory as memories of particular nodes are available to the application. In this case, if particular processes of an application consume only the memory of the size they need, page swapping may be unnecessary. Consequently, the application running in a multi node environment may result in time shorter than $t(1)/N$.
2. Another example of when superlinear speed-up may occur is related to applications in which processing time may be dependent on particular input data. For example, if an application is to find a particular argument, element or a solution that meets a certain criterion in a search space, it is not known in advance how much time it will take. Specifically,

a code running on 1 computing device may need to go through almost the whole search space before finding a solution. On the other hand, on 2 or more computing devices, one of processes or threads may find the correct solution much sooner than in $t(1)/N$ time. Obviously, this assumes that the application terminates as soon as the solution has been found rather than having finished browsing the whole search space. The former may be enough if it can be determined that a solution is good enough. If, however, the best solution has to be found and it is unknown in advance then the whole search space needs to be browsed.

Parallel efficiency can be defined as follows:

$$pe(N) = \frac{t(1)}{N \cdot t(N)} \tag{3.3}$$

where:

- $t(1)$ – execution time of the sequential application running on a system with 1 computing device,
- $t(N)$ – execution time of the application on a system with N computing devices.

$pe(N)$ close to 1 corresponds to very good speed-up, close to ideal.

Cost of computations can be defined as $C = t(N) \cdot N$.

3.2 MASTER-SLAVE

Master-slave is a programming paradigm that can be applied to many practical algorithms for parallelization. Specifically, the paradigm distinguishes two types of actors:

1. Master – partitions input data into data packets which are sent to slaves for processing. Upon receiving results of processing data packets, these are merged in order to obtain a final solution to a problem.
2. Slave – responsible for processing of received data packets and returning results. A slave is not aware of existence of other slaves and just communicates with the master.

The master-slave paradigm may be implemented in various ways. First, let us consider the scenario shown in Figure 3.1. The master divides input data into a number of data packets equal to the number of slaves. The number of slaves may in turn be equal to the number of computing devices e.g. CPU cores that are available apart from one dedicated to the master. The master then executes the following steps (Figures 3.2 and 3.3):

1. Send data packets to particular slaves.

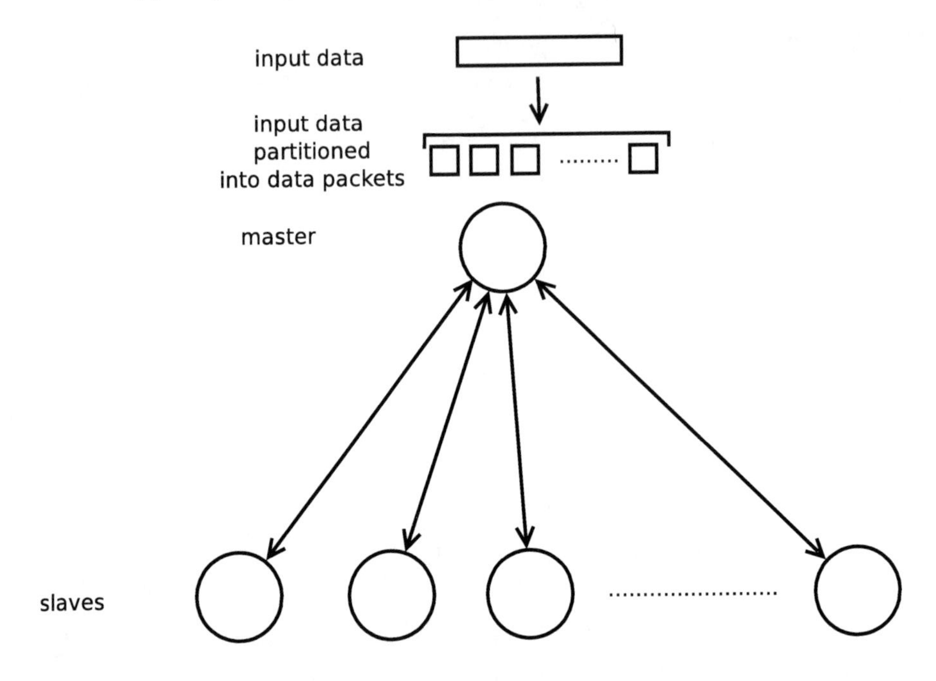

FIGURE 3.1 Basic master-slave structure with input data

2. Wait and receive results from slaves.

3. Combine results received from slaves into a final result.

If we assume slaves running on computing devices of same speeds then this approach allows practically perfect load balancing across computing devices although distribution of different data packets may be delayed for slaves with higher ids. Assuming that times of computations are considerably larger than communication times, such a scenario allows high speed-ups.

It should be noted that the master can merge results received from slaves after it has received all the subresults. Alternatively, the master could merge each individual subresult into a current result right after it has received the subresult. The aforementioned scenario has the following drawbacks, though:

1. If computing devices are of various speeds then processing will not be balanced and the execution time of the application will be determined by the slowest device.

2. If processing of various data packets takes various amounts of time, again processing will not be balanced and the execution time of the application will be determined by processing of the data packet that takes the largest time. In general, it may not be possible to predict how much time processing of an individual data packet will take. Processing may

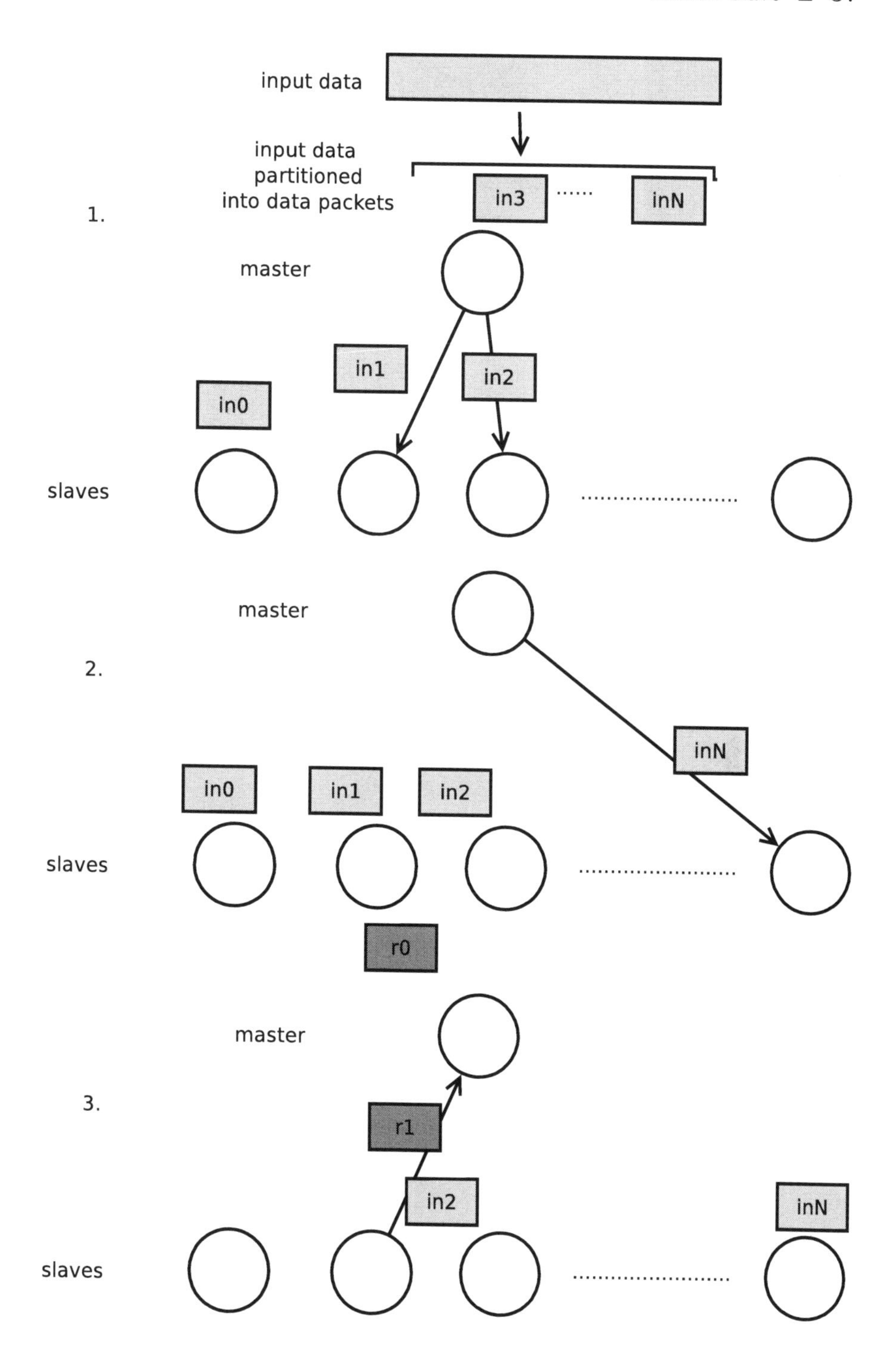

FIGURE 3.2 Flow of the basic master-slave application over time, 1 of 2

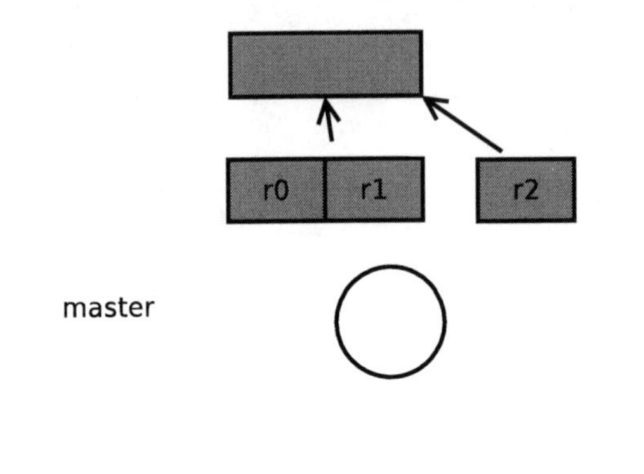

4.

inN

slaves

final result

master

5.

slaves

FIGURE 3.3 Flow of the basic master-slave application over time, 2 of 2

depend not only on the input data size but also on the data itself. For example, integration of a function over a given range using an adaptive quadrature [170, 38] algorithm will depend not only on the length of the range but mainly on the function within the range. Analysis of a position on a chess board will heavily depend on the position of pieces on the board, not only on the number of the pieces.

Consequently, another approach can be adopted as shown in Figures 3.4 and 3.5. Specifically, input data is divided into a larger number of data packets. This allows such granularity so that computations can be balanced across computing devices even if processing of various data packets takes various amounts of time. Obviously, good load balancing will be possible if while processing a data packet that takes much time, other computing devices can be made busy processing other data packets. Practically, there should be at least a few times more data packets than the number of computing devices. It should be noted, though, that increasing the number of data packets considerably will fine tune load balancing but at the cost of handling many data packets. As can be seen from Section 3.1.2, sending a data packet involves startup time. In case of too many data packets, this will lead to a large overall overhead. Consequently, given a certain input data size there will be a trade-off between load balancing and the overhead leading to a sweet spot for the number of data packets. This is further discussed with an experiment using a parallel application in Section 6.2.

While using more data packets allows for a more dynamic code that can handle computing devices of various speeds and data packets that require different amounts of time to process, the solution can be improved further. From the point of view of a slave, it waits for a new data packet, processes the data, sends a result back and repeats this procedure. It is clearly visible, that there is idle time on the computing device used by the slave between when the latter finished computations and starting computations on a yet to be received new data packet. This scenario is shown in Figure 3.6a.

A solution to improve this method is to make sure that a slave has enough data to process after it has finished processing a data packet. This requires sending a new data packet in advance such that it is already available for a slave when it finished processing the previous data packet. This solution is shown in Figures 3.6b and Figures 3.7-3.8. Ideally, it would eliminate idle time on the slave side. However, since processing of a data packet may take only a short time, it may not be enough. Sending more data packets to a slave in advance would eliminate this risk but at the cost of potential load imbalance in case processing of some data packets sent in advance would require much time to process.

3.3 GEOMETRIC SPMD

This paradigm mainly refers to modeling and solving phenomena that have natural, geometric representation in a real world. Several examples can be given involving:

1. distribution of heat/time evolution of heat [28],

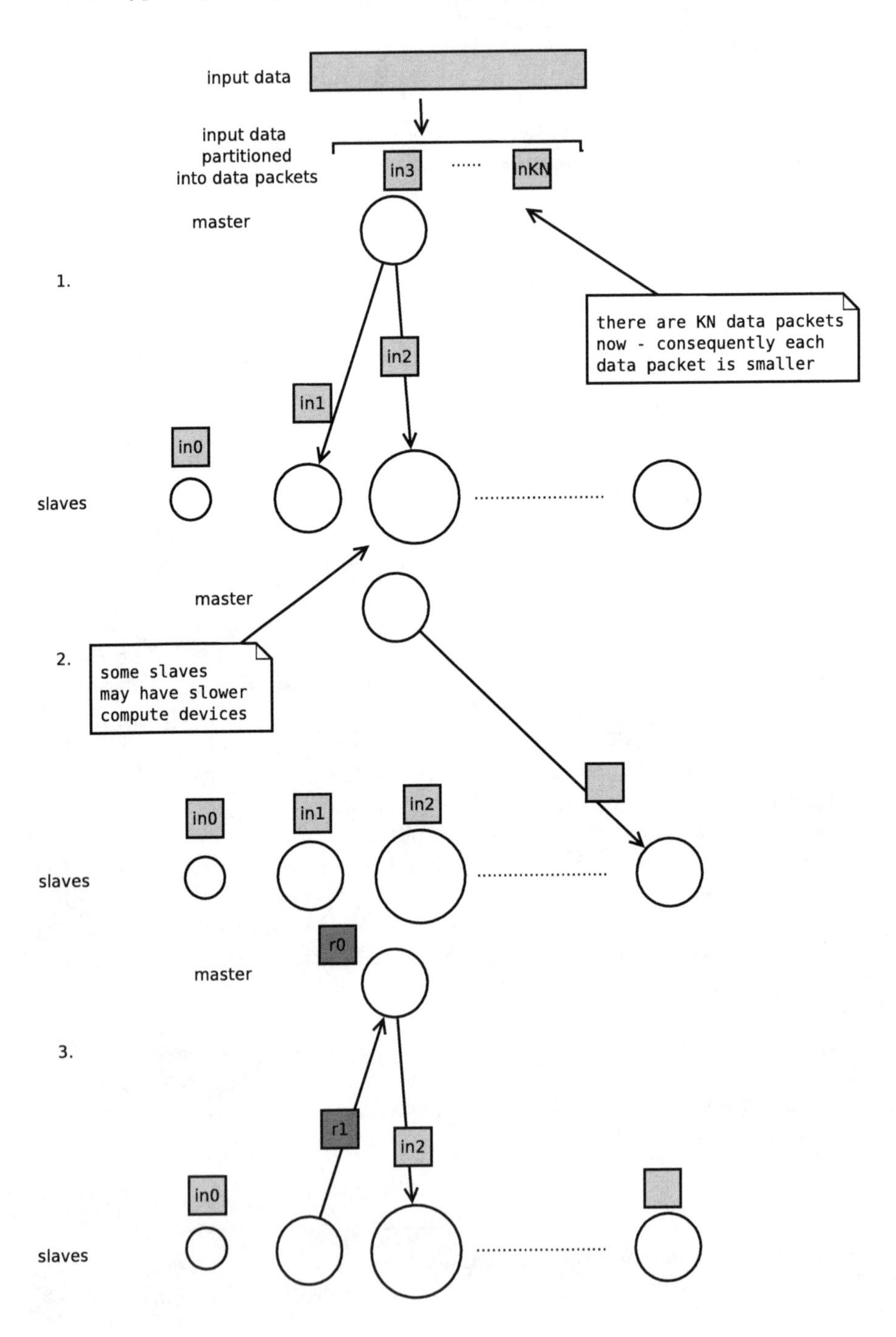

FIGURE 3.4 Flow of the basic master-slave application with more data packets over time, diameters denote execution times, 1 of 2

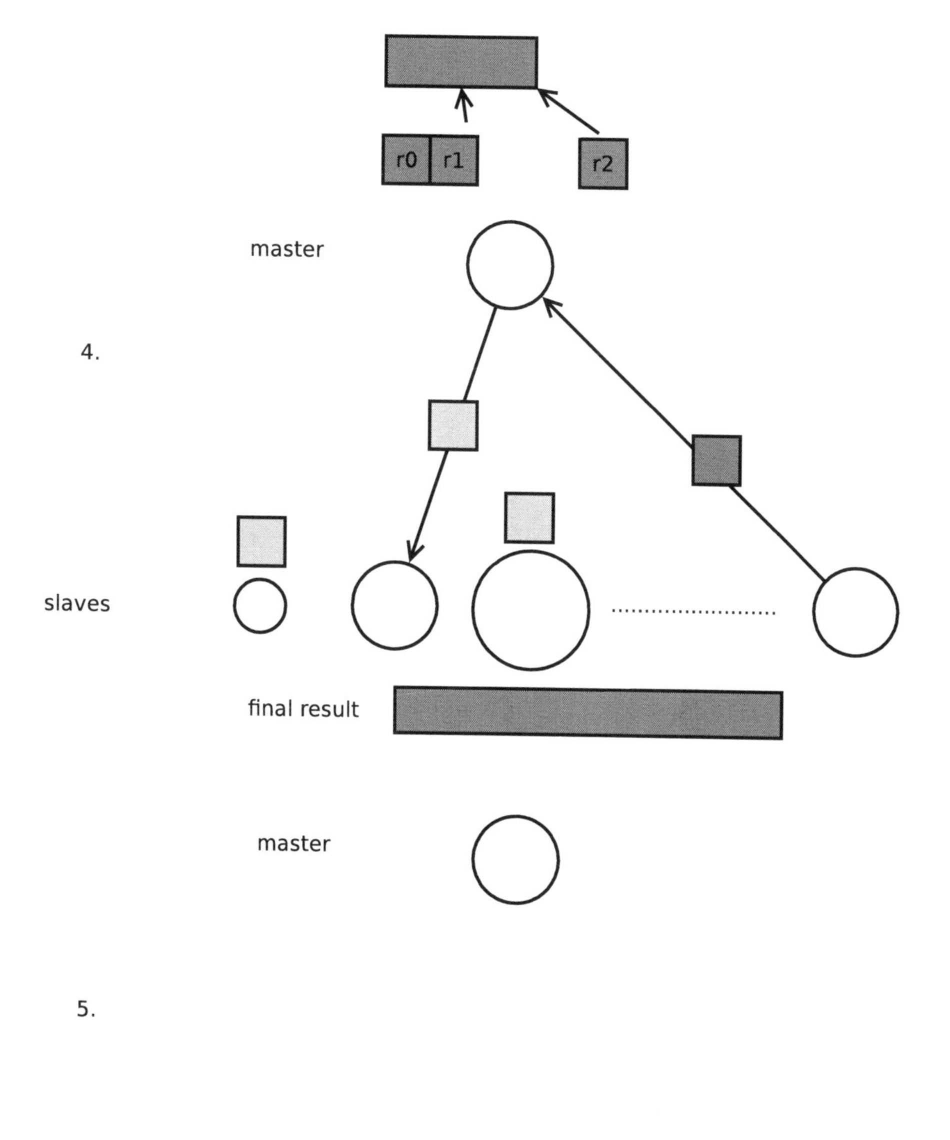

FIGURE 3.5 Flow of the basic master-slave application with more data packets over time, diameters denote execution times, 2 of 2

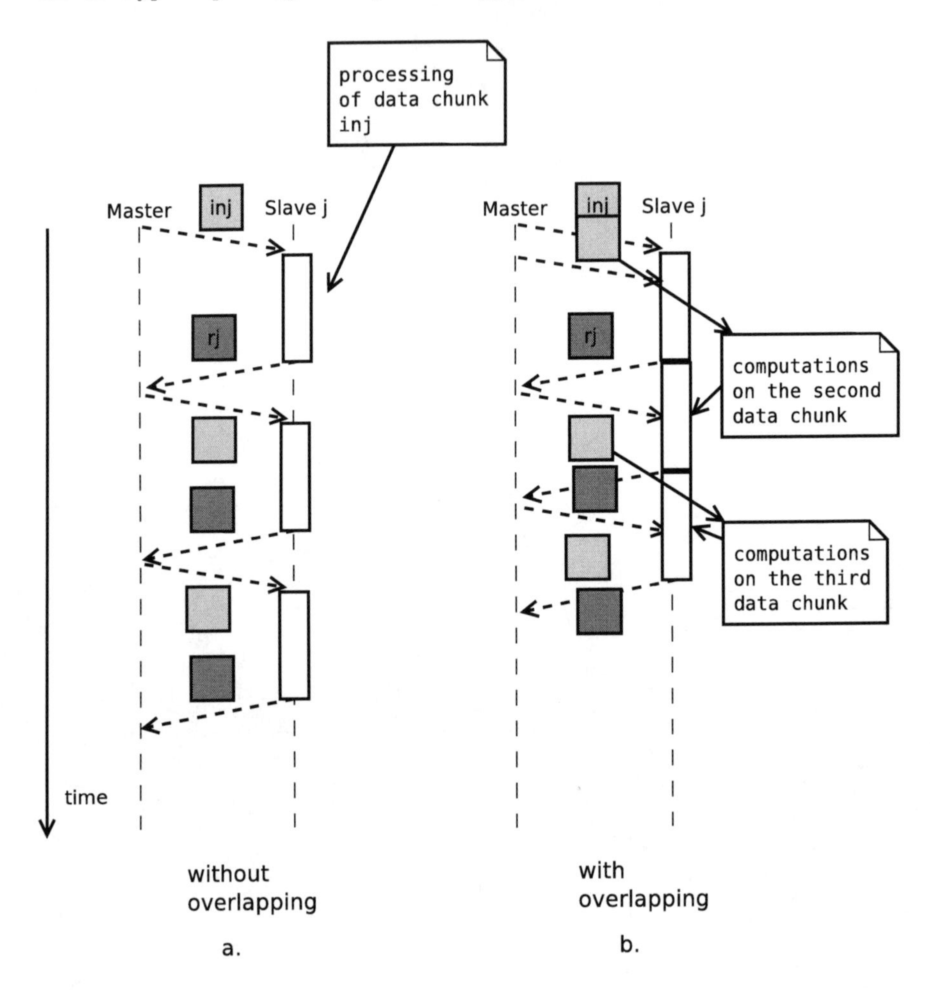

FIGURE 3.6 Interaction diagram for master-slave processing with overlapping communication and computations

2. electromagnetics [100, 148, 147, 112],
3. weather prediction [27],
4. phenomena in medicine [50] etc.

Such problems use a geometric representation of a domain, such as in two or three dimensions. The domain is then considered as a grid of cells each of which typically refers to a physical location or part of space in which a given phenomenon is simulated. Then the following will need to be defined for a specific problem:

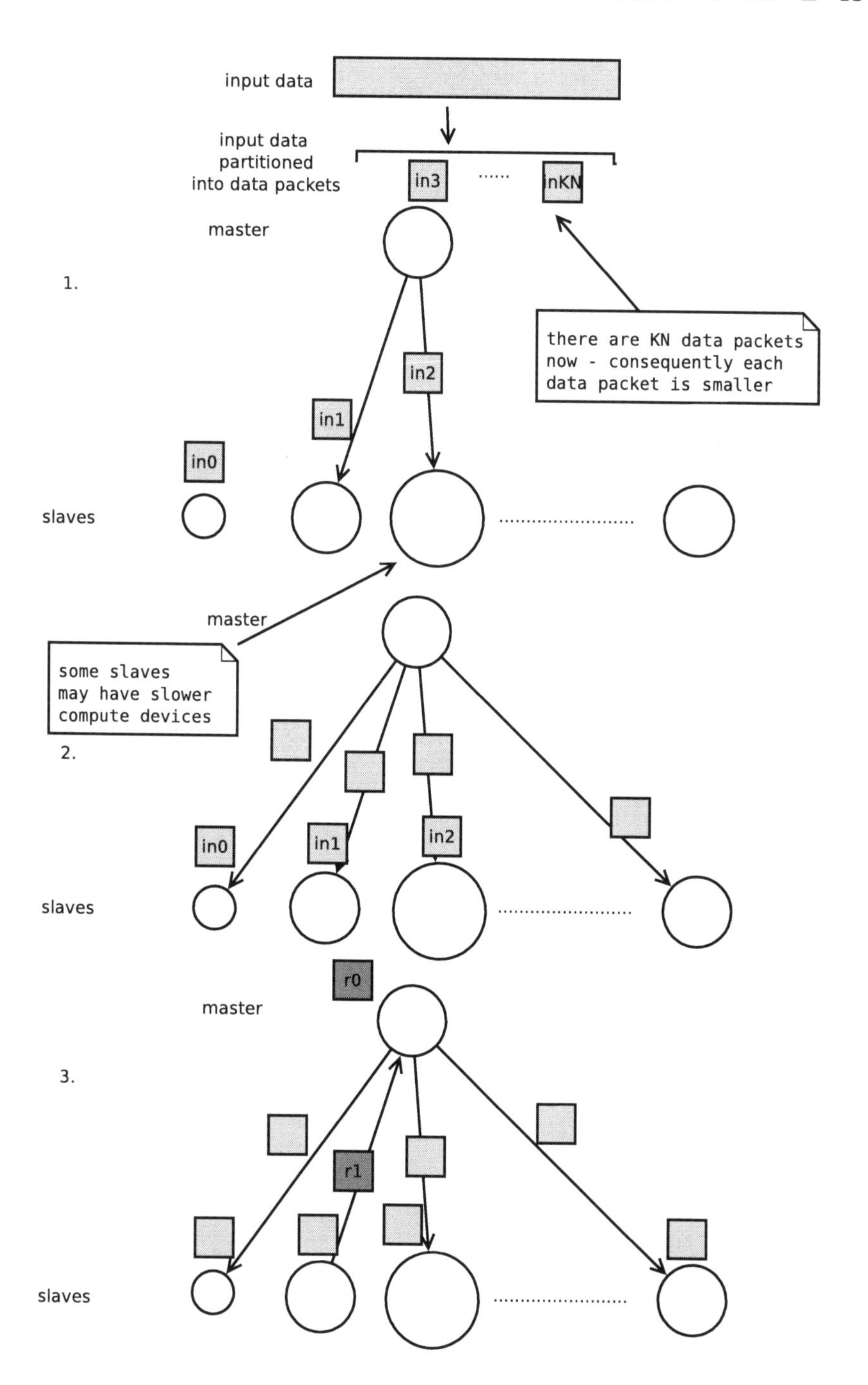

FIGURE 3.7 Flow of the basic master-slave application with more data packets over time and overlapping communication and computations, diameters denote execution times, 1 of 2

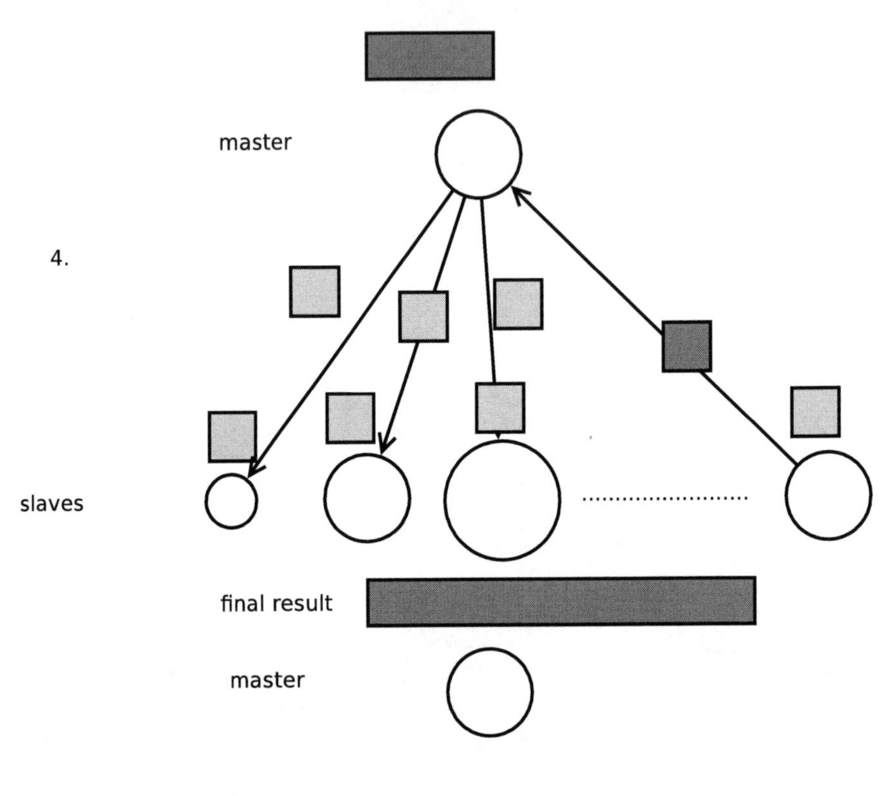

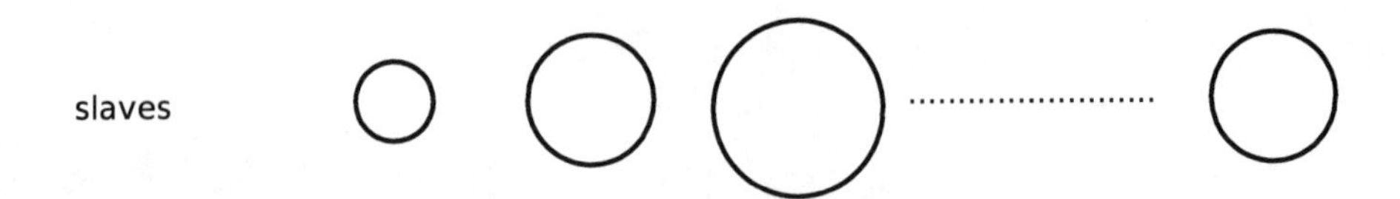

FIGURE 3.8 Flow of the basic master-slave application with more data packets over time and overlapping communication and computations, diameters denote execution times, 2 of 2

1. Assignment of proper variables to each cell. Such variables will be problem specific and may represent e.g. temperature, pressure, wind direction or speed for weather prediction, electromagnetic field vectors for elec-

tromagnetic simulations [148], intracellular calcium ion concentration in medical simulations [50] etc.

2. Specification of proper update equations for each cell with dependency on values computed in a previous time step for the same and neighboring cells.

A simulation will then proceed in successive iterations that will typically correspond to time steps. When solving the system of linear equations using the Jacobi or related methods, iterations will usually allow divergence to a solution. In each iteration, cells throughout the whole domain will need to be updated. In a parallel implementation, the whole domain will be partitioned into disjoint subdomains (with additionally so-called ghost cells described below) each of which will be assigned to a distinct process or thread for updates. An example of such partitioning for a two dimensional domain is shown in Figure 3.9.

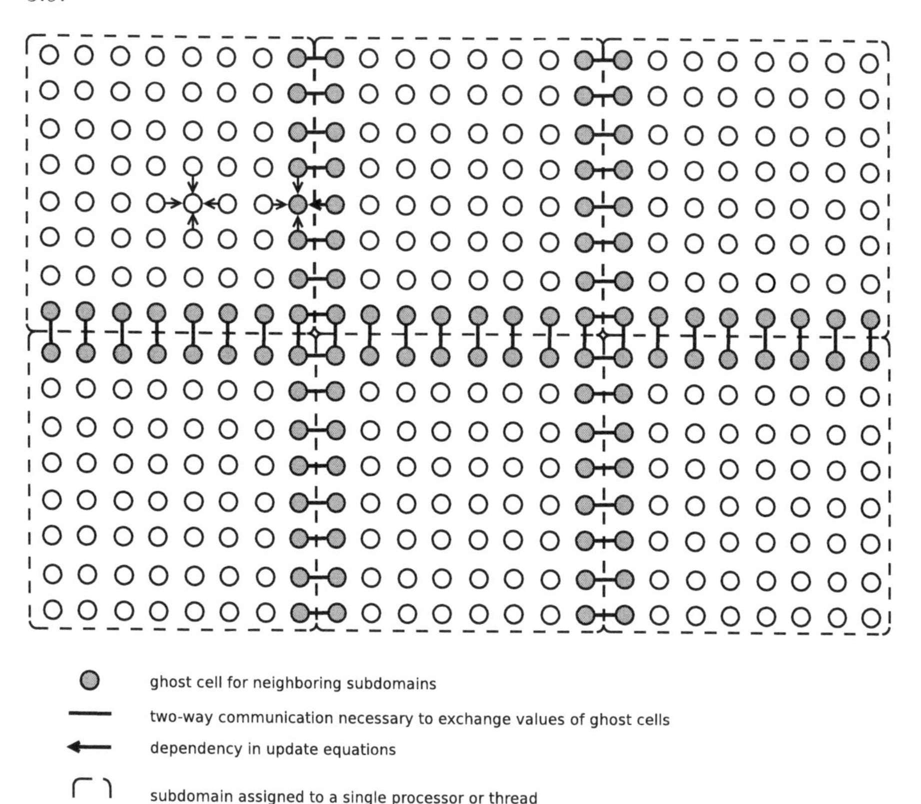

FIGURE 3.9 Partitioning of a 2D space into subdomains

After a process or thread has updated all its cells, the next iteration should follow. Before this can be done, however, processes or threads handling neighboring subdomains will need to exchange values of boundary cells that will be used by their neighbors for updates of their boundary cells in the next iteration. Such cells are called ghost cells. Consequently, in order to reach good performance and high speed-ups of the parallel code the following will need to be performed at the same time:

1. balancing computations among parallel processes or threads,
2. minimization of communication times for sending and receiving ghost cells.

Communication times will depend on the underlying communication infrastructure. Proper partitioning should minimize the number of bytes exchanged and the number of neighbors in the communication, especially between processes on various nodes. As an example, for rectangular domains the following partitioning strategy can be applied (Figure 3.10):

1. Partition the domain with planes parallel to XY, YZ, and XZ planes respectively.
2. There are planes parallel to YZ, XZ and XY planes respectively. The former planes cut the domain into equally sized subdomains. The domain is of size X, Y and Z which is input data to the partitioning algorithm.
3. The goal of the partitioning strategy is to balance computations while minimizing communication between subdomains. Since the planes cut the domain into equally sized parts, computations are balanced assuming each cell in every subdomain requires the same computational effort. However, taking into account possibly various sizes of the domain: X, Y and Z it is necessary to find such numbers of planes cutting the domain in each dimension that communication is minimized. In a 3D space, subdomains in the middle of the domain have six neighbors. The total number of cells that need to be exchanged, from the point of view of a single subdomain, is as follows:

$$2(\frac{X}{x+1}\frac{Y}{y+1}+\frac{X}{x+1}\frac{Z}{z+1}+\frac{Y}{y+1}\frac{Z}{z+1}) \tag{3.4}$$

Assuming that the number of processes P is given this forces the following condition

$$(x+1)(y+1)(z+1)=P \tag{3.5}$$

X, Y and Z are given. The goal would become to find such x, y and z that the value given by Equation 3.4 is minimized. This leads to minimization of

$$\frac{X}{x+1}\frac{Y}{y+1}\frac{z+1}{z+1}+\frac{X}{x+1}\frac{Z}{z+1}\frac{y+1}{y+1}+\frac{Y}{y+1}\frac{Z}{z+1}\frac{x+1}{x+1}= \quad (3.6)$$

$$\frac{1}{P}(XY(z+1)\frac{Z}{Z}+XZ(y+1)\frac{Y}{Y}+YZ(x+1)\frac{X}{X})= \quad (3.7)$$

$$\frac{XYZ}{P}(\frac{z+1}{Z}+\frac{y+1}{Y}+\frac{x+1}{X}) \quad (3.8)$$

Finally, the goal is to find such x, y and z that $(x+1)(y+1)(z+1) = P$ and $\frac{z}{Z}+\frac{y}{Y}+\frac{x}{X}$ is minimized.

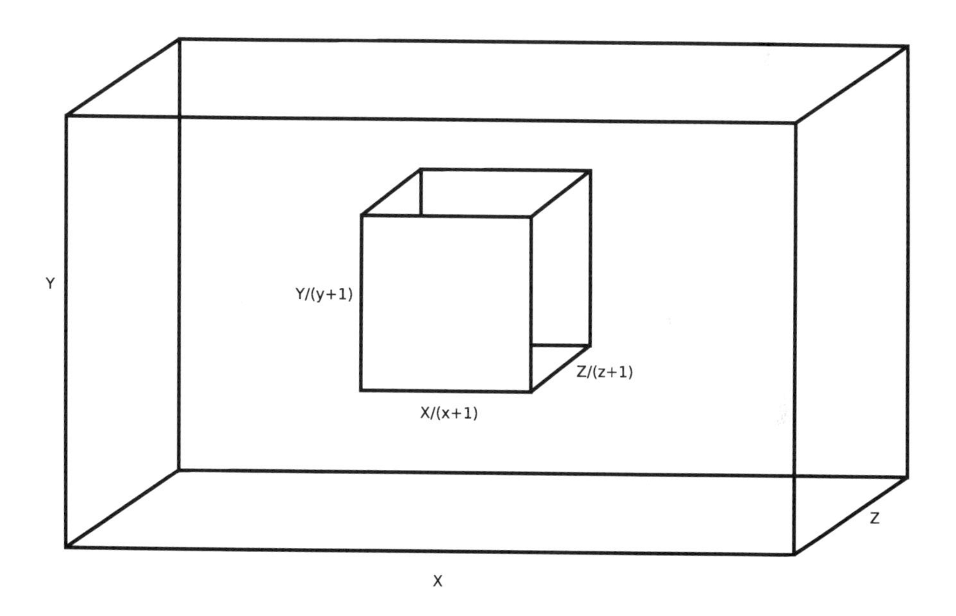

FIGURE 3.10 Subdomain within the whole input data domain in 3D

So far we have considered a problem in which all cells throughout the domain require the same computational effort. In case various cells have different computational weights that result in different update times then it needs to be taken into account during partitioning of the domain into subdomains. There are several approaches that can be adopted for load balancing. For simplicity, in this case, the so-called Recursive Coordinate Bisection (RCB) is demonstrated. Figure 3.11 presents the same domain as considered before but with various cell weights. In a real world scenario, this may correspond to a simulation of a phenomenon with a source and distribution of waves from the source throughout space. Other equations or accuracy might be preferred closer to the source than in the other part of space. A recursive partitioning algorithm can work as follows:

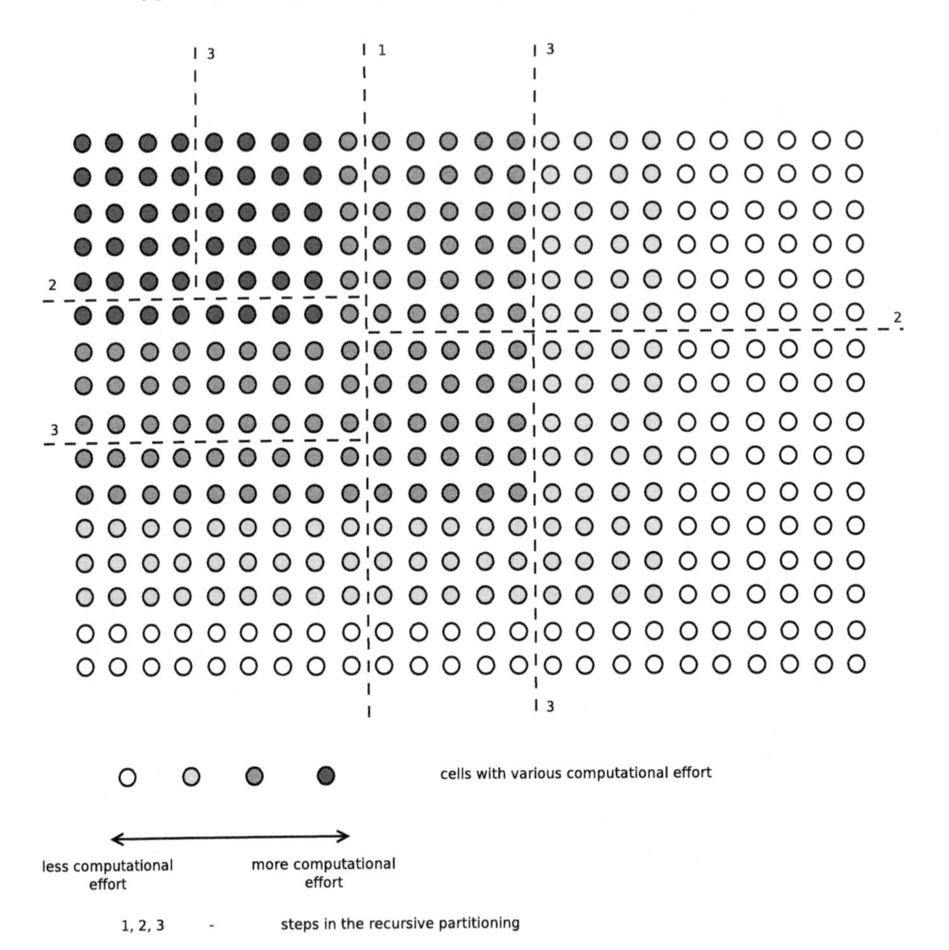

FIGURE 3.11 Partitioning of a 2D space with various cell weights into subdomains

1. Partition the current domain (initially the whole domain) into two subdomains by cutting it with a plane such that the dimension with the largest width is cut by a perpendicular plane. In two dimensions, as shown in Figure 3.11, the algorithm cuts with lines.

2. Apply step 1. to the resulting subdomains until a required number of subdomains is reached.

Then there comes the main simulation loop. As indicated above, the basic steps of the simulation are outlined in Figure 3.12 with computations and communication performed in every iteration. One of disadvantages of this approach is that communication follows the whole computational step.

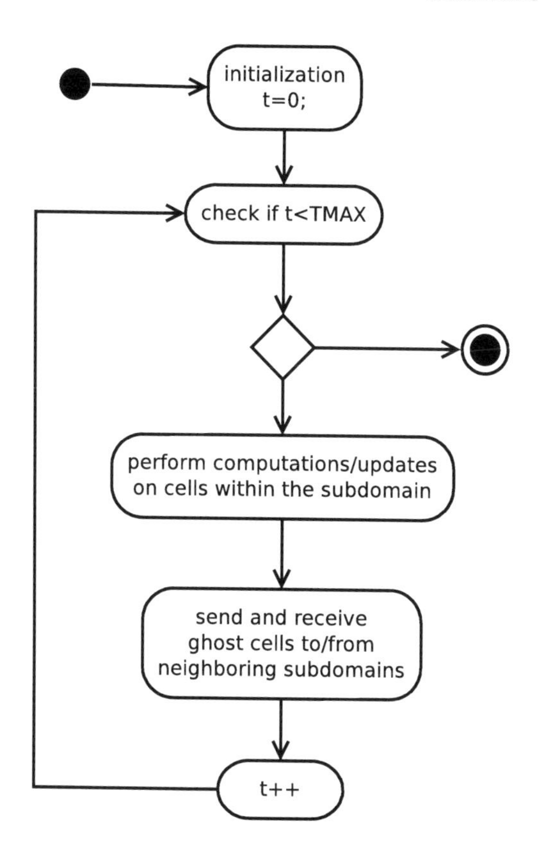

FIGURE 3.12 Activity diagram with basic steps of a typical parallel geometric SPMD application

It is typically possible to improve this approach by introducing so-called overlapping computations and communication described in more detail in Section 6.1. In this particular case, as shown in Figure 3.13, each process or thread can perform the following steps:

1. Update its boundary cells first. It can be done assuming that values for ghost cells from the previous iteration are available.

2. Start sending the updated boundary cells to processes or threads handling neighboring subdomains. At the same time, start receiving updated values for ghost cells.

3. Update all remaining cells within the domain i.e. the interior cells. It should be noted that the communication started in step 2 and these updates can potentially be performed at the same time thus saving some time compared to the solution shown in Figure 3.12.

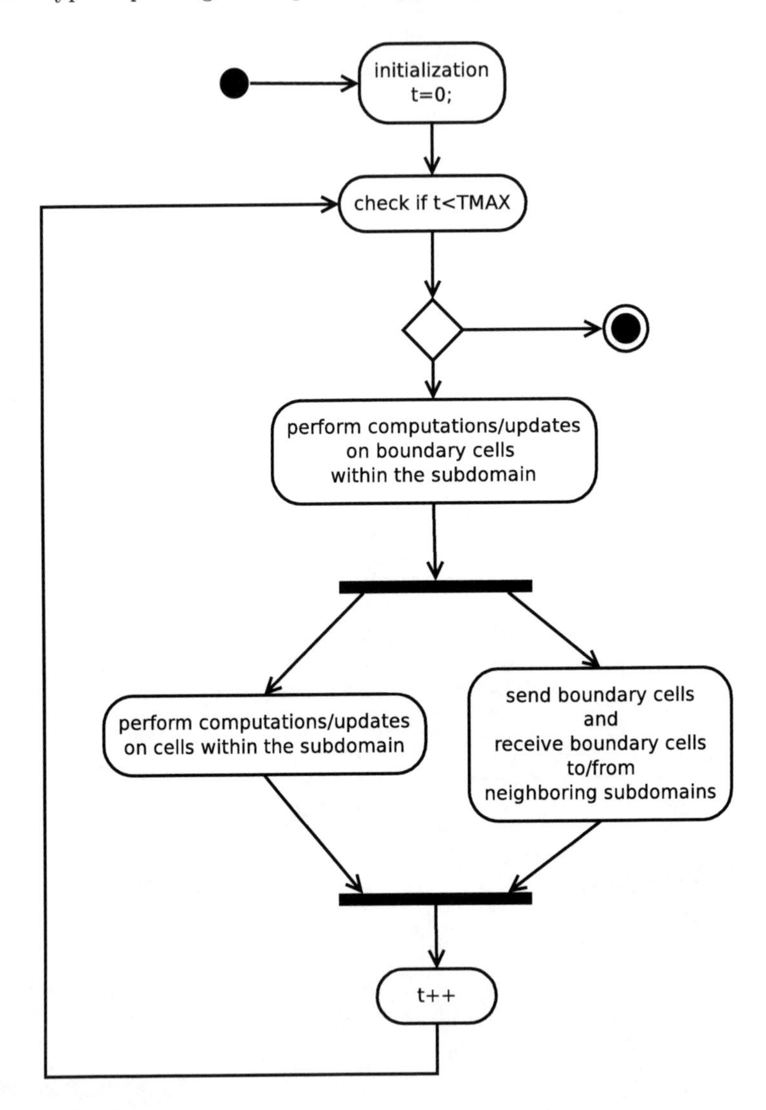

FIGURE 3.13 Activity diagram with basic steps of a typical parallel geometric SPMD application with overlapping communication and computations

4. Upon completion of the updates, wait for completion of the communication started in step 2. This is necessary before proceeding with the following iteration.

The aforementioned solution assumed that in the main simulation loop each process or thread maintains its load throughout successive iterations. In some problems, however, computational load changes in successive iterations. This may lead to considerable differences in weights among cells. In such a case, a process or thread with the largest execution time per iteration will become a bottleneck. Consequently, a dynamic repartitioning strategy may need to be applied in order to balance computational load again. Figure 3.14 presents a general activity diagram with the main simulation loop and dynamic load balancing within the loop. The following should be taken into account, however:

1. checking load of neighbors takes time,
2. load balancing time can be very large.

A decision needs to be taken if it is beneficial to perform load balancing. The load balancing scheme shown in Figure 3.15 works as follows:

1. Check whether to perform a check for the need for repartitioning in the given iteration. Checking every specified number of iterations is a simple solution but is not able to adapt to various rates of load changes in various applications. A solution could be to adapt the checking frequency based on the rate of repartitioning. Such a solution is presented next.
2. Check imbalance among the node itself and its neighbors. If imbalance exceeds a certain threshold then dynamic load balancing could be invoked. However, this process can also consider:
    - the time for repartitioning,
    - the remaining number of iterations.

The scheme shown in Figure 3.15 compares if time lost from imbalance is larger than time required for repartitioning and based on this assessment starts dynamic load balancing or not. Figure 3.16 presents the same scheme but improved with dynamic adjustment of the following parameters:

1. Dynamic adjustment of checking frequency. This is based on the frequency of dynamic load balancing actually performed. If dynamic load balancing is performed every time it is checked for then `load_balancing_step_check` is decreased. On the other hand, if checking is too frequent and actual load balancing is performed rarely then `load_balancing_step_check` can be increased. As a result, the overhead within the main simulation loop is smaller. Specifically, there is variable `load_balancing_last_time_counter` that is initialized to 0. Every time dynamic load balancing is actually performed it is increased by 1, otherwise decreased by 1. If it is greater than a certain positive threshold then the step for when to check for load balancing is decreased by a factor of 2. If it is lower than a certain negative threshold then the step for when to check for load balancing is increased by a factor of 2.

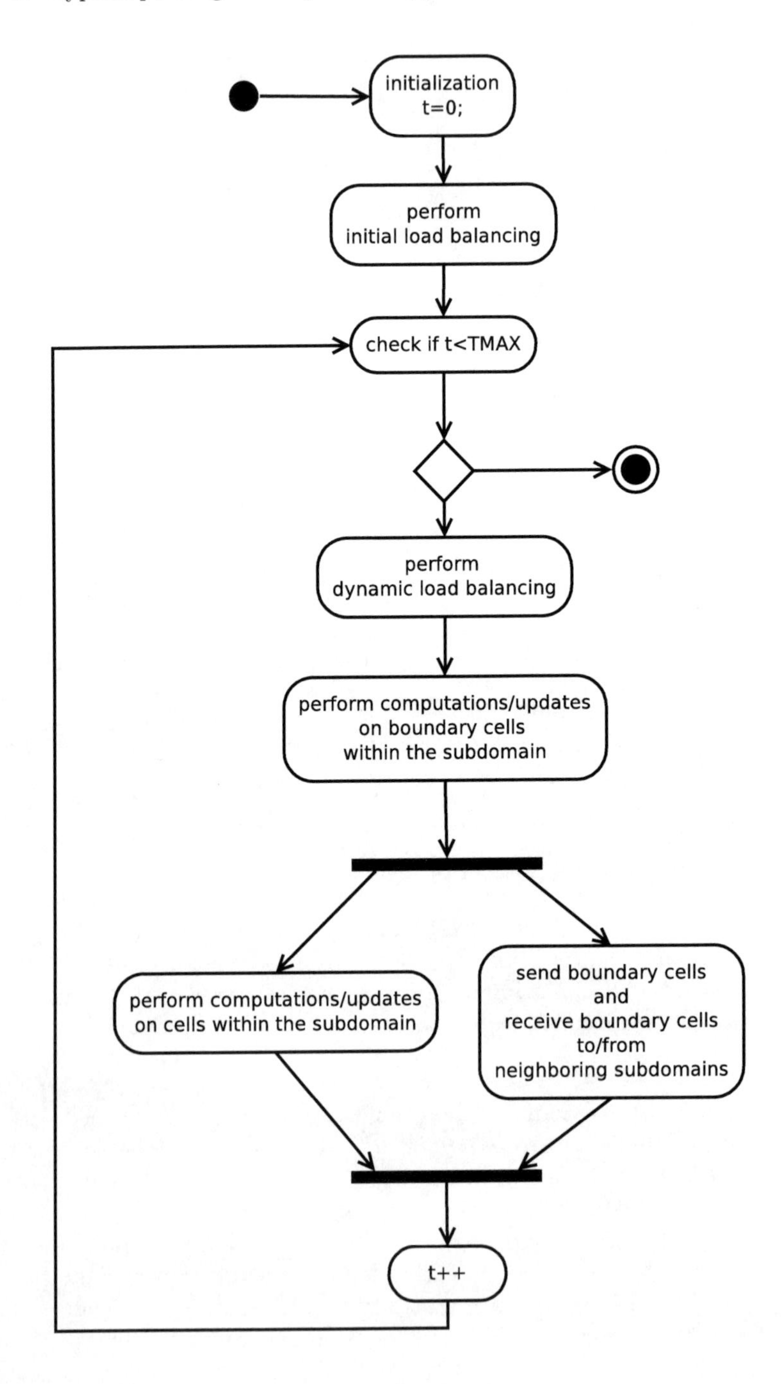

FIGURE 3.14 Activity diagram with basic steps of a typical parallel geometric SPMD application with overlapping communication and computations and dynamic load balancing

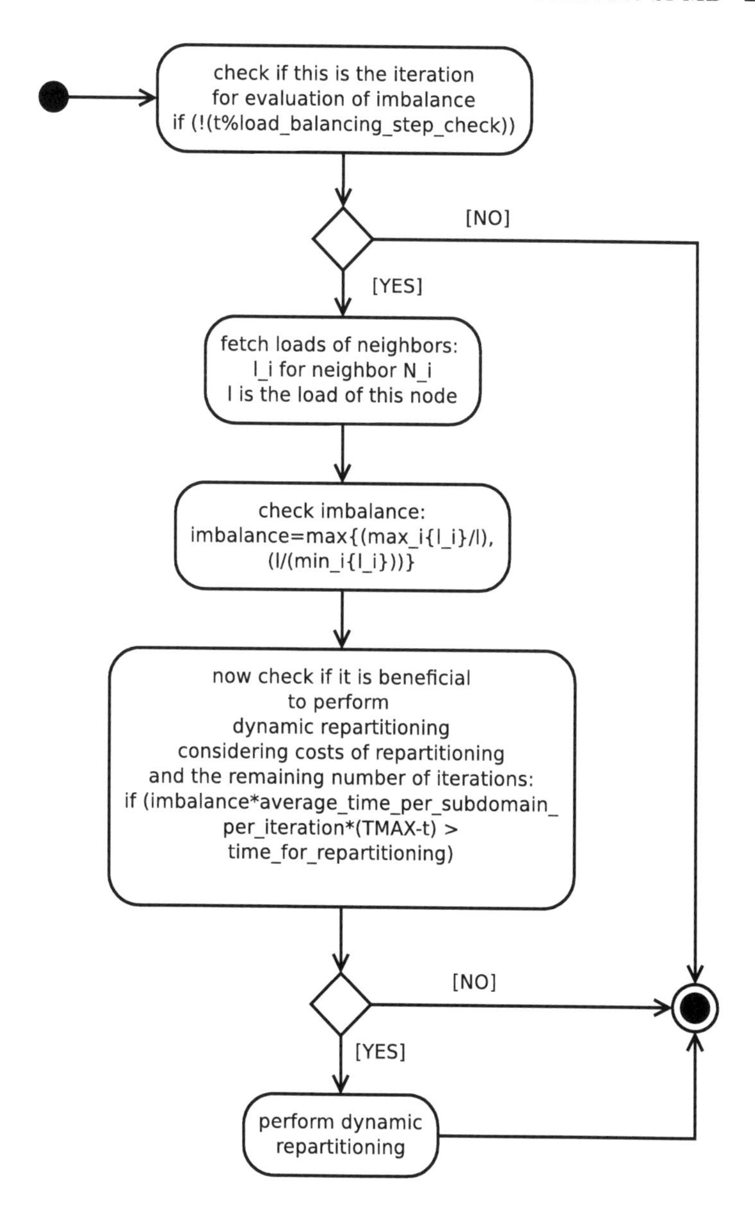

FIGURE 3.15 Activity diagram for the dynamic load balancing step

2. Dynamic assessment of repartitioning times. Repartitioning time may depend on changes of load within the domain, domain size etc. Since it is a crucial value in the algorithm, precise assessment is necessary. One of the possible solutions is a running average of previous times (such as two or more times) which adjusts to the problem.

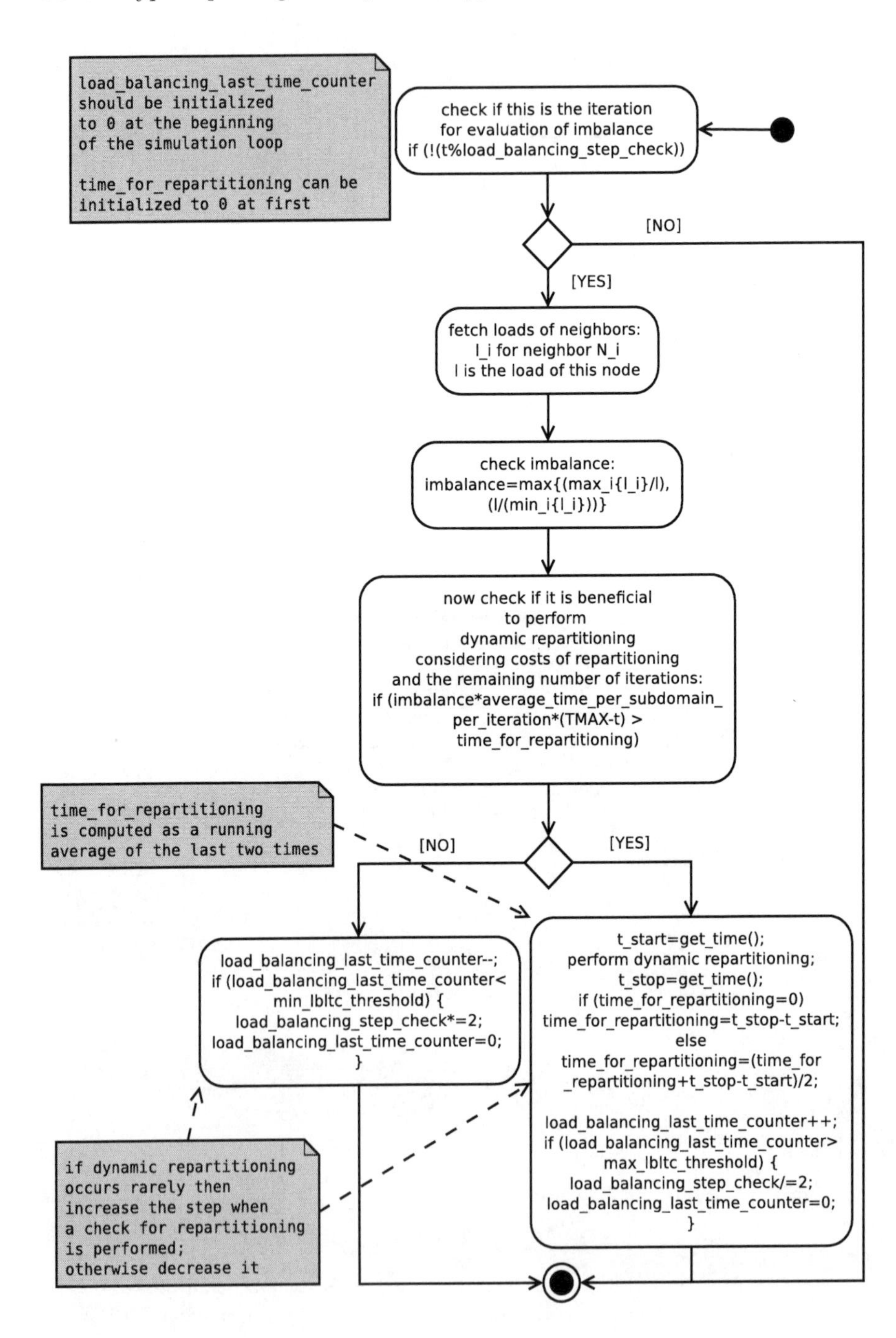

FIGURE 3.16 Activity diagram for the dynamic load balancing step improved with dynamic load balancing step adjustment

The aforementioned scheme requires fetching information about loads of nodes handling neighboring subdomains. This process in itself can be a costly operation. One of the following two solutions can be adopted in order to minimize this overhead:

1. engaging a separate thread for fetching load information in the background,

2. adding load information to the data (values of ghost cells) that is sent between neighbors anyway. This adds only a small number of bytes to the messages sent anyway and saves on startup time.

3.4 PIPELINING

In pipeline type processing, one can distinguish distinct stages of processing through which each of the input data packet needs to go. As shown in Figure 3.17, input data is considered as a stream of input data packets that enter a pipeline. Each of the pipeline stages is associated with code that can be executed on a separate core, processor or accelerator. Specifically, some codes might preferably be executed on such a computing device that is especially well suited for execution of such code. This might be a rationale for selection of this type of processing. However, such pipeline processing can suffer from performance bottlenecks if some pipeline stages process data packets much longer than the others.

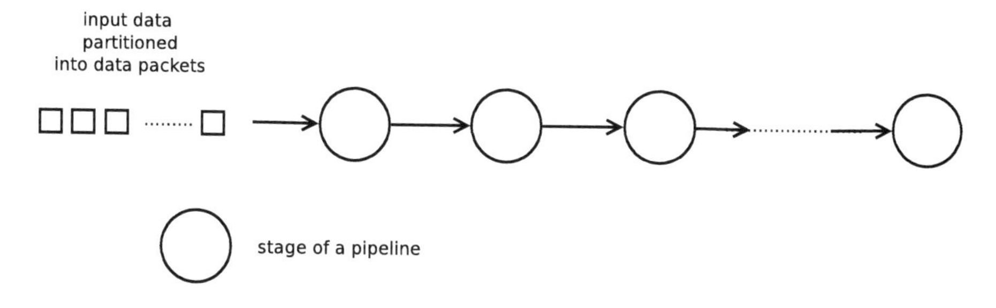

FIGURE 3.17 Basic structure for pipelined processing

It should be noted that depending on timing in processing and communication in and between particular stages of the pipeline, that a prior stage may already send its output to a following node even if the latter is still processing its own data packet (Figure 3.18). This means that when a following node finished processing a data packet, it may already have another ready for being processed.

Pipelining seems to be the paradigm useful in scenarios in which specific stages of the pipeline are most efficiently executed on a specific node or a processor, much better compared to others.

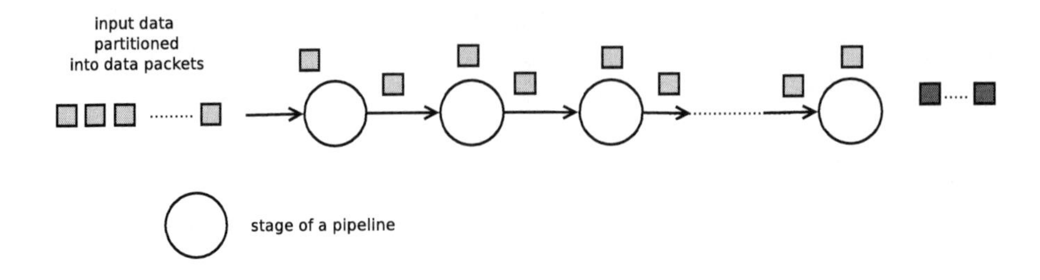

FIGURE 3.18 Basic flow for pipelined processing

3.5 DIVIDE-AND-CONQUER

Divide-and-conquer is a technique that is based on the following steps:

1. If there is a direct solution to a given problem then it is applied directly.
2. If there is no direct solution then the problem is partitioned into subproblems. For each of the subproblems, the approach is repeated i.e. steps 1. and 2. are performed.

A divide-and-conquer algorithm can be represented by a tree in which nodes denote subproblems (Figure 3.19). The tree can be characterized in terms of the following:

1. Degrees of the nodes. In some cases a problem is divided into a certain fixed number of subproblems. For instance, in quick sort an array is always divided into two subarrays: one with elements smaller and the second with elements larger than a pivot. On the other hand, in alpha beta search in chess a current problem corresponding to a certain position on the chessboard will be divided into the number of subproblems equal to the number of legal moves for the player to move. Numbers of legal moves will most likely be different for various positions.
2. Depth of the tree. In some algorithms the depth may be the same for all branches of the tree. For instance, the first search using an alpha beta algorithm may be limited until a predefined depth is reached. Only then, based on obtained results, deepening of certain searches can be performed. It should be noted that in an initial phase of the game, there may be enough moves to reach the depth in all the branches. Later in the game, however, an algorithm may reach a mate or stalemate during the search. This means that the tree will not then be balanced.

Figure 3.19 shows an imbalanced tree with various node degrees and depths for various branches of the tree. In terms of computations that are to be performed within nodes, various configurations can be distinguished as well, considering:

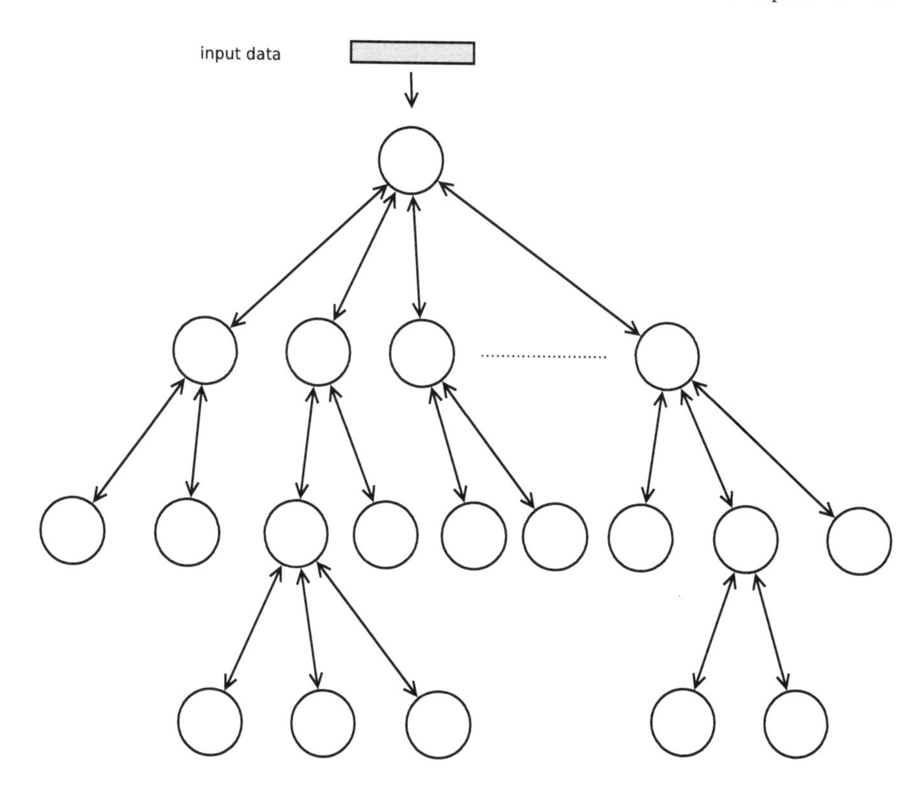

FIGURE 3.19 Imbalanced tree corresponding to the divide-and-conquer processing paradigm

1. Uniformity of computations within leaves. In some applications all leaves perform exactly the same computations. In merge sort, each leaf of the tree checks which of two numbers is larger and puts these in correct order. In alpha beta search each leaf of the tree is associated with a board position that needs to be evaluated. In chess there might be a mate or stalemate in which case the evaluation completes quickly. Otherwise, normal evaluation based on board position and pieces is performed which takes more time.

2. Ratio of computations performed within leaves to the time required to pass input and output data between tree nodes. In merge sort this ratio is small while in the case of complex evaluation in chess it is considerably larger.

By definition of divide-and-conquer, intermediate (non-leave) nodes partition input data that come from nodes closer to the root. On the way back, these tree nodes are to aggregate subresults from nodes closer to leaves and pass intermediate results to nodes closer to the root.

In alpha beta search in chess evaluation of a position performed in leaves may consider not only the pieces each player has on the board but also positional advantages, immediate threats etc. This may be a time consuming process. There may be a trade-off between how accurate evaluation is and the size of the tree to be analyzed within a timeout for a move. On the other hand, in merge sort processing in leaves only requires comparison and potentially exchange of numbers which is fast. Next, merging in intermediate nodes takes more and more time as arrays to be merged are larger and larger.

Let us first consider a balanced tree with the same node degrees equal to 2. Such a tree can be found in algorithms such as mergesort. If we assume that pairs of processes/threads can communicate/synchronize at the same time then computations can be organized as shown in Figure 3.20. Particular processes or threads and required communication or synchronization is shown. It should be noted that in this allocation, some processes or threads are responsible for nodes of the tree at various levels. Flow of computations, on the other hand, is shown in Figures 3.21 and 3.22.

It should be noted that, in general, a divide-and-conquer application can be highly dynamic in terms of the following:

1. Generation of subtrees may depend on previously computed results. For instance, in alpha beta search subtrees in subsequent branches of the tree may be cut off depending on thresholds obtained in previous branches.

2. Time required to process various nodes, especially leaves of the tree may vary as noted above.

This makes parallelization of such a scenario much more difficult. Efficient processing requires balancing computations among available nodes. As a consequence of point 2 above, processing of various subtrees may take various amounts of time and at least in some cases it can not be predicted in advance. As an example, in alpha-beta search in chess it can not be done as it depends on how a particular position will unfold and the processing tree may depend on previously obtained values of alpha and beta. It should also be noted that, depending on the problem, a tree corresponding to a solution may have many more leaves than the number of processes or threads that could be launched and run efficiently in a real parallel system. Consequently, each of processes or threads will need to process subtrees rather than deal with individual leaves. In general, this is good for parallel efficiency because there is a good chance that the ratio of computations to communication is large. It should be noted, however, that processing of a node clearly requires prior provision of input data. Subsequently, upon processing of a node, its results are necessary in order to merge with results of other nodes at the same level. This means that communication time will need to be taken into account in order to compute

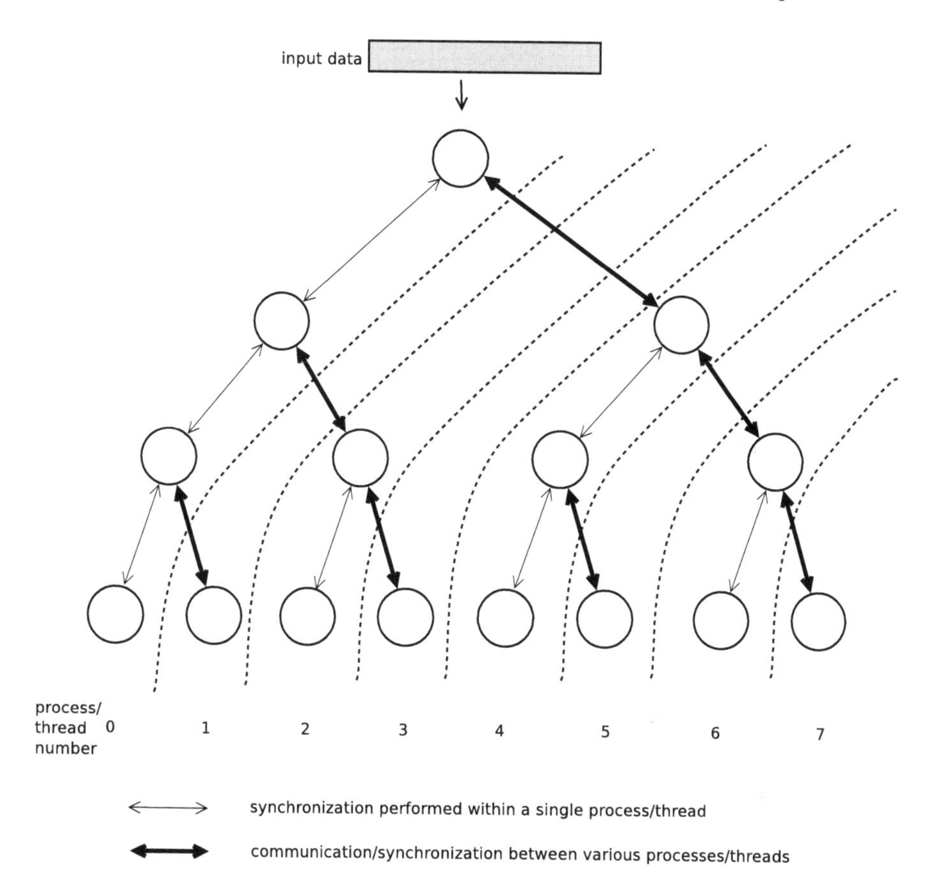

FIGURE 3.20 Sample partitioning of the divide-and-conquer tree

the total running time of the algorithm. Figure 3.23 presents such a general divide-and-conquer tree.

In order to parallelize a general divide-and-conquer tree efficiently, a dynamic approach with partitioning of the tree can be used [38, 47]. Specifically, if there are several processing nodes then, in terms of load balancing, each of the processes or threads can perform steps depicted in Figure 3.24.

These steps can be outlined as follows for a process or thread on each processing node:

1. If there is at least one subtree available then fetch and process one.
2. If there is no data to process then request a subtree from one of the neighbors of the processing core/node. There may be various load balancing algorithms that determine:
    - which neighbors can be considered for load balancing,

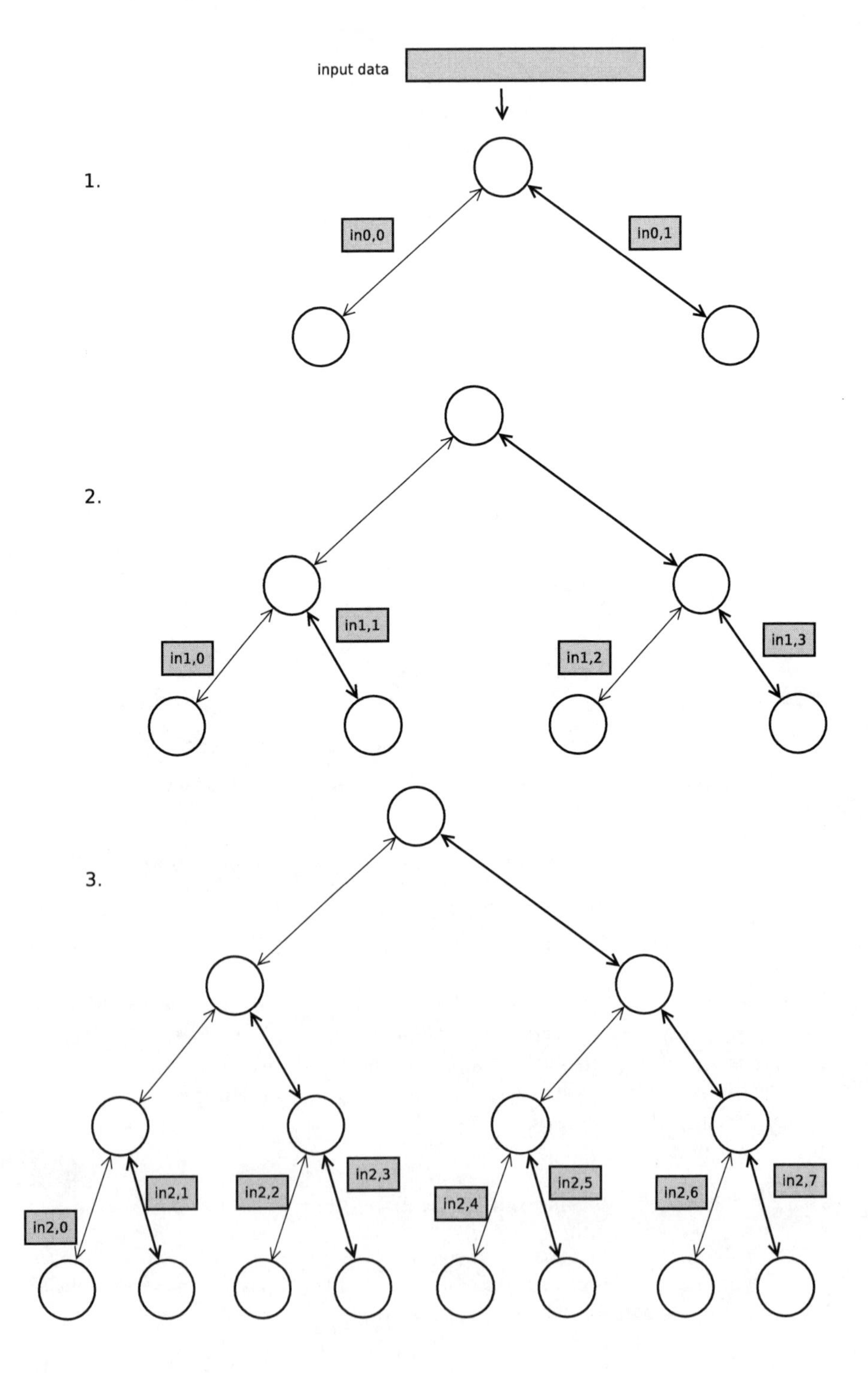

FIGURE 3.21 Flow in the divide-and-conquer application, 1 of 2

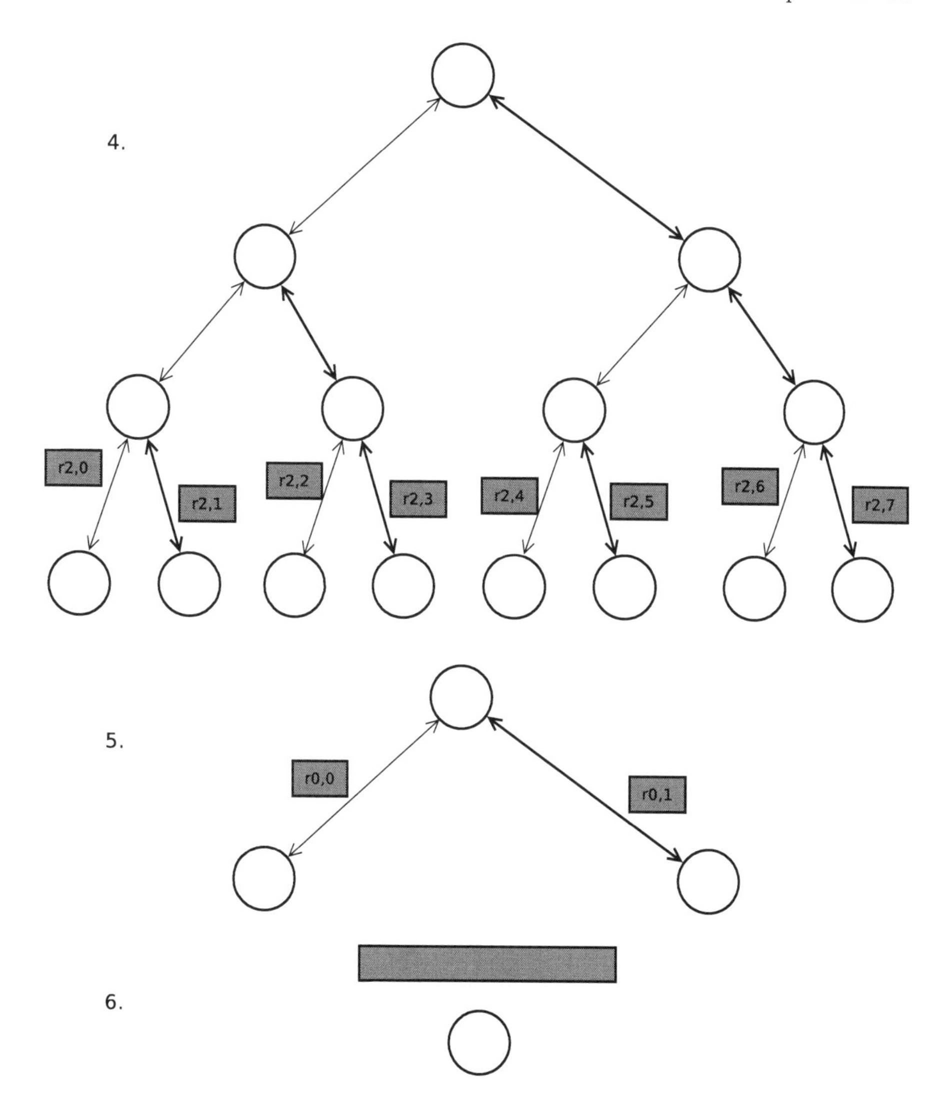

FIGURE 3.22 Flow in the divide-and-conquer application, 2 of 2

- how much data should be fetched from neighbors – how many subtrees in this case.

This algorithm allows cutting off subtrees from nodes that have large trees available. It should be noted that a subtree can be cut off and be sent for processing to another process or thread. This idea is related to concepts of *work stealing* and *work sharing* used in scheduling of multithreaded computations. Work stealing uses a mechanism in which underloaded processors request and

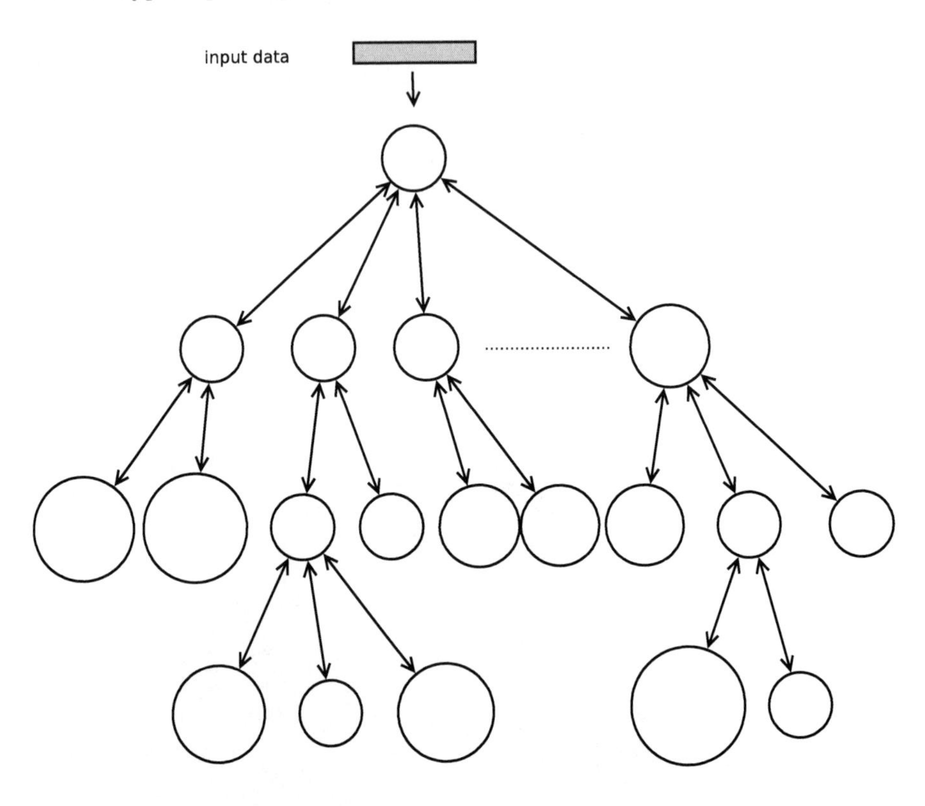

FIGURE 3.23 Imbalanced tree corresponding to the divide-and-conquer processing paradigm with various node processing times

fetch work from overloaded ones and work sharing uses spawning work on processors and migration such that load is balanced. In this case work (a subtree) is generated on demand. Processing of distinct subtrees is independent. However, a process or thread that is to merge results will need to wait for results from all subtrees at a given level. Consequently, the following performance related issues arise:

1. There may be idle time between when an idle node requests a new subtree and the time data is available. One solution to cope with this is data prefetching described in Section 6.1. Specifically, before an available subtree is to be processed, a request may be sent in the background for sending another subtree. However, it should be noted that prefetching too many subtrees may overload the given node because quite often the real computational effort for a subtree can not be predicted in advance [38].

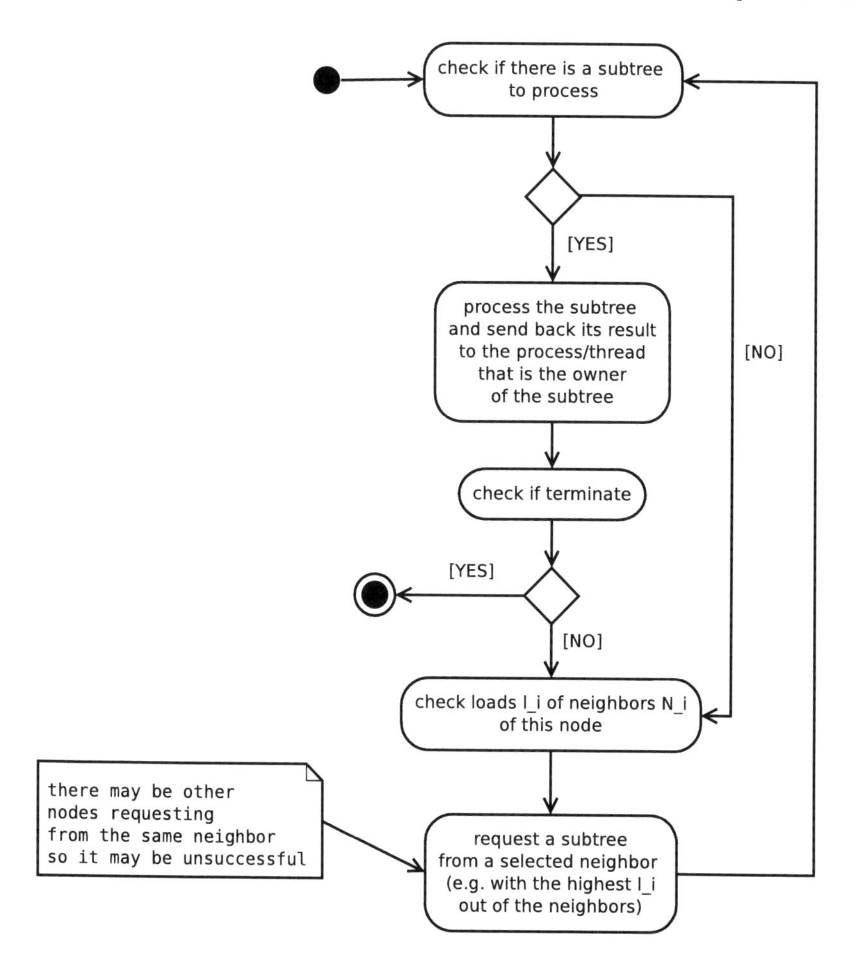

FIGURE 3.24 Basic activity diagram of a process/thread with computations and load balancing for divide-and-conquer processing

2. Determination of a computational effort required to process a given subtree. This may be difficult in general but may be dependent on the predicted depth of the subtree. For instance, if processing of a subtree with depth d is scheduled then the average size of the tree may be estimated as $a \cdot avnd^d + b$ where $avnd$ is the average node degree in the subtree which may also be estimated in advance and a and b denote coefficients.

An improved activity diagram is shown in Figure 3.25. In this case, two threads are distinguished:

1. data processing thread that is responsible for fetching subtrees from a queue or a buffer,

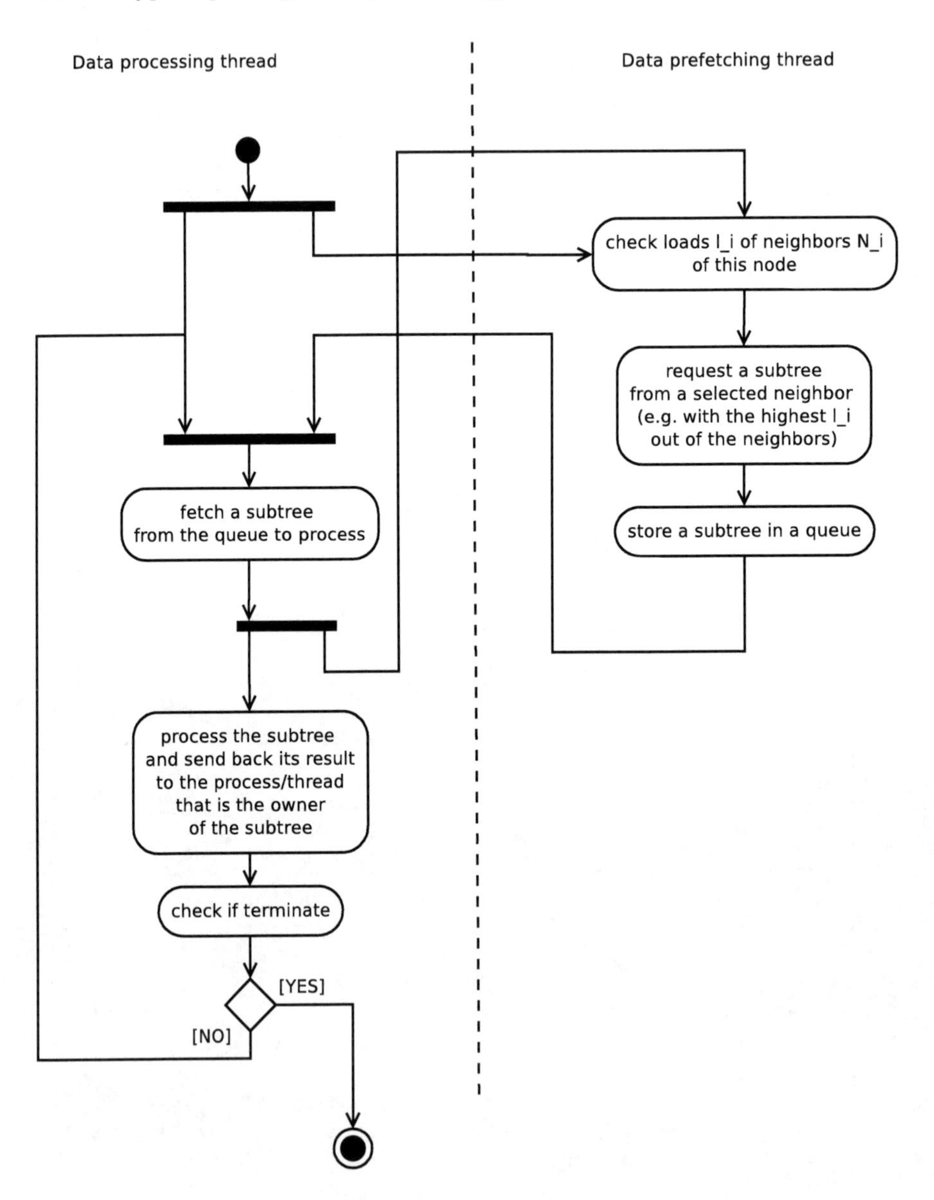

FIGURE 3.25 Improved activity diagram with computations and load balancing performed by two threads for divide-and-conquer processing

2. data prefetching thread that prefetches subtrees to be processed by the first thread.

It should be noted that firstly a subtree is requested and is processed immediately. Before processing, though, a new subtree is requested so that it can actually be prefetched during the processing time of the first subtree. This technique actually implements overlapping communication and computations which is described further in Section 6.1. Note that processing of the next subtree starts only after it has been fetched and processing of a previous subtree has been finalized. These steps follow until a termination flag is set.

If a process or thread that owns a subtree receives a request from another process or thread to partition its subtree then it can cut off a subtree close to the root following the concept introduced in [38]. This potentially results in cutting off a subtree that potentially (but not always) requires a reasonably large processing time. Then such a subtree needs to be stored and be sent to the process or thread that has requested work. In many cases, the subtree will actually be represented by its root and additional values such as the required depth of the subtree for which computations are required. For instance, in alpha beta search in chess a subtree can be represented by all of the following:

1. position on the board (pieces and their locations),
2. the player to move,
3. information about castling for the players,
4. information about previous moves (for checking for repeated positions).

Figure 3.26 presents a divide-and-conquer processing tree and cutting off a subtree with a root node S1. The subtree was cut off when there was a request from another process or thread at the time the owner of the tree was processing the node marked with R1. It can be seen that there was a possibility to cut off a subtree as close to the root as possible. Next, as shown in Figure 3.27, if there is another request that arrives when the process or thread is processing the node marked with R2, a new subtree can be spawned with a root marked with S2. This time there is no possibility to spawn a subtree at the level closest to the root. In general, such a scheme may consider the highest level from the root or the smallest depth of a subtree to be cut off so that computational effort required to process a subtree is significant compared to the time required to send out the subtree and receive a corresponding result. Furthermore, the scheme requires a proper load balancing algorithm that would determine the following:

1. which nodes are considered as neighbors to a given node,
2. how requests from many neighboring nodes are handled – which ones are accepted and which are rejected,
3. what thresholds are used in order to determine if a node is a candidate to request subtrees from.

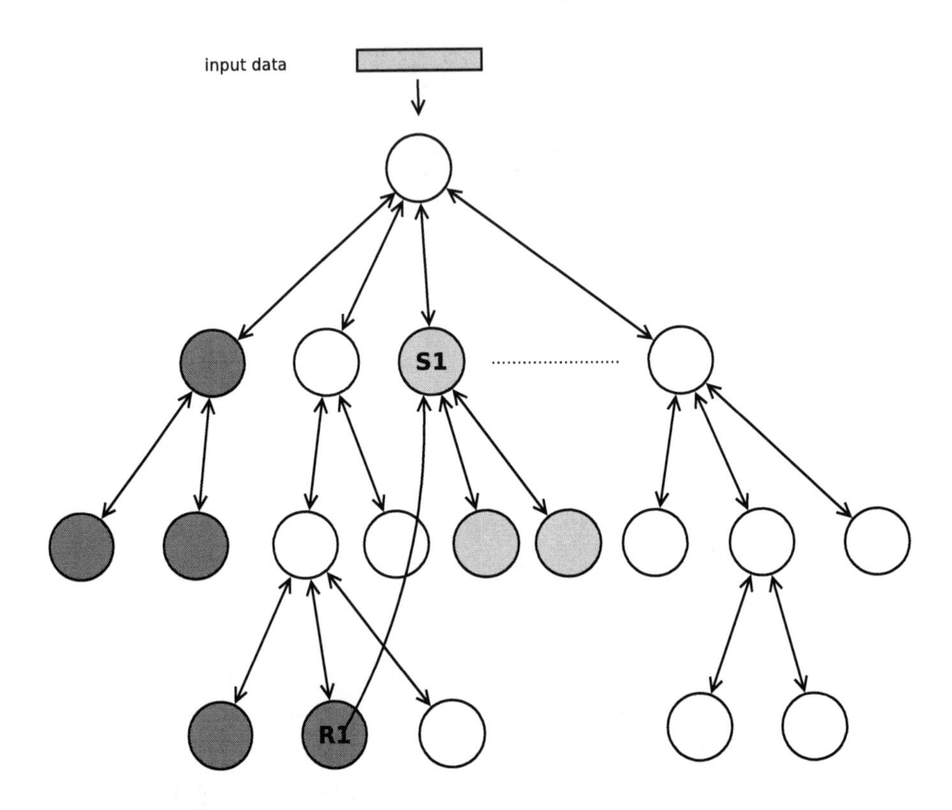

FIGURE 3.26 Cutting off subtrees for divide-and-conquer processing

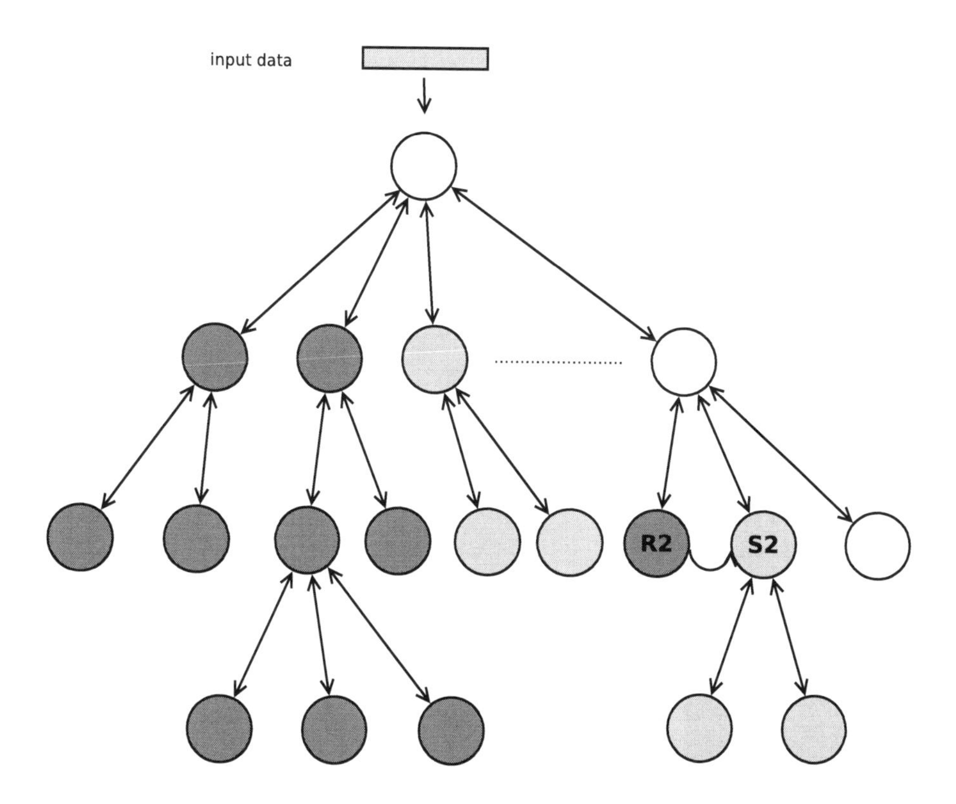

FIGURE 3.27 Cutting off subtrees for divide-and-conquer processing – continued

CHAPTER 4

Selected APIs for parallel programming

CONTENTS

This chapter aims to introduce of some of the most useful parts of popular APIs for parallel programming in today's modern computing systems. Description of the APIs is meant to introduce functions and methods most often used in solving practical problems. The presented APIs are representative in terms of types of systems on which these are used for programming parallel applications. The APIs include:

- Message Passing Interface (MPI) – suitable for cluster systems as it allows interprocess communication among processes running on various cluster nodes as well as between processes within each node.
- OpenMP – for multithreaded parallel programming suitable for shared memory systems, also includes directives for offloading computations to e.g. coprocessors. OpenMP typically allows easy extension of a sequential program to a parallel program using directives and library functions.
- Pthreads – for multithreaded parallel programming suitable for shared memory systems. Includes mechanisms such as mutexes and conditional variables that allow a thread to sleep and be notified in response to data availability from other threads.
- CUDA – for programming NVIDIA GPUs. An application started on a CPU launches parallel processing by calling kernel functions that are executed on a GPU by many lightweight threads in parallel.
- OpenCL – for programming parallel computing devices, including GPUs and multicore CPUs. An application started on a CPU launches parallel processing by calling kernel functions that are executed on a computing device (OpenCL distinguishes the concept of a *Compute Device*) by many work items in parallel.
- OpenACC – for parallel applications to be offloaded to accelerators, typically GPUs. An application, similarly to OpenMP, includes directives and library calls that refer to parallelization.

In their respective contexts, key areas of the APIs are presented with the most focus on MPI, OpenMP, CUDA and OpenCL. Selection is based on required elements for implementation of master-slave, geometric SPMD and divide-and-conquer parallelism, presented in Chapter 5, as well as useful and modern elements of the APIs available in the latest versions. The most important of the presented elements in the APIs are summarized in Tables 4.1 and 4.2 and include:

- programming model and application structure – that define whether a parallel application consists of processes and/or threads and how these interact which each other,
- data exchange and flow/data synchronization and consistency – how data is exchanged and synchronized between processes/threads and how views of threads can be synchronized,
- sample application – an example using the given API,
- additional features – useful features available in the API, not necessarily used in a typical application using the API.

TABLE 4.1 Presented features of parallel programming APIs, 1 of 2

	Programming model and application structure	Data exchange and flow/data synchronization and consistency	Sample application	Additional features
MPI	Sections 4.1.1-4.1.3, dynamic process creation – Section 4.1.15	data exchange and synchronization through sending messages or one-sided communication – Sections 4.1.4-4.1.12	Section 4.1.13	multithreading – Section 4.1.14, parallel I/O – Section 4.1.16
OpenMP	Sections 4.2.1-4.2.3, 4.2.5	thread synchronization, data synchronization and consistency – Section 4.2.4	Section 4.2.6	SIMD directives – Section 4.2.7, offloading – Section 4.2.8, tasking – Section 4.2.9
Pthreads	Section 4.3.1	mutual exclusion – Section 4.3.2, barriers – Section 4.3.4, synchronization and consistency – Section 4.3.5	Section 4.3.6	condition variables – Section 4.3.3

TABLE 4.2 Presented features of parallel programming APIs, 2 of 2

	Programming model and application structure	Data exchange and flow/data synchronization and consistency	Sample application	Additional features
CUDA	Sections 4.4.1, 4.4.3	synchronization of threads within a group, atomic operations – 4.4.2, synchronization using streams – Section 4.4.5	Section 4.4.4	streams and asynchronous operations – Section 4.4.5, dynamic parallelism – Section 4.4.6, Unified Memory – Section 4.4.7, management of GPU devices – Section 4.4.8, using shared memory – Section 4.4.4
OpenCL	Sections 4.5.1-4.5.2	queuing data reads/writes and kernel execution – Section 4.5.3, synchronization – Section 4.5.4	Section 4.5.5	using local memory – example in Section 4.5.5
OpenACC	Sections 4.6.1-4.6.2	data management and synchronization – Section 4.6.3	Section 4.6.4	asynchronous processing and synchronization – Section 4.6.5, device management – Section 4.6.6

4.1 MESSAGE PASSING INTERFACE (MPI)

4.1.1 Programming model and application structure

Message Passing Interface (MPI) [6] has been the de facto standard for parallel programming for several years already. The programming model used by MPI is well suited for distributed memory systems such as clusters that are composed of computing nodes coupled with a fast interconnect. A standard MPI application is depicted in Figure 4.1 and consists of processes that perform computations and interact with each other by sending messages. Each process of the application is identified with a so-called rank, unique within a group of processes in an MPI communicator. `MPI_COMM_WORLD` denotes a communicator encapsulating all processes started in an MPI application.

MPI 1 draft was presented at the Supercomputing conference in 1993 which was followed by MPI 1.1 in June 1995. MPI 2 added functions such as one-sided communication, spawning processes dynamically and parallel I/O, described in this book in Sections 4.1.12, 4.1.15 and 4.1.16 respectively. MPI 3.0 was released in September 2012 and included non-blocking collective operations and extensions to the one-sided API within MPI. MPI 3.1 from June 2015 included, in particular, non-blocking I/O function specifications.

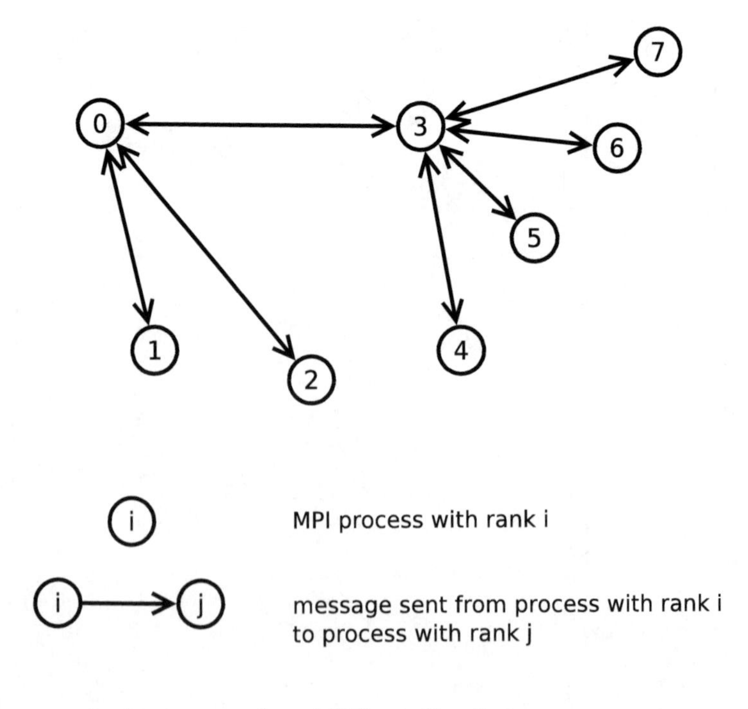

FIGURE 4.1 Architecture of an MPI application

4.1.2 The world of MPI processes and threads

The basic component of an MPI application is a process. A process executes an algorithm expressed in the code making use of resources such as memory, caches, disk etc. and communicates with other processes in order to send, receive data and synchronize. The following characterize the world of MPI processes within an application:

1. Each process within a group is identified by a so-called rank which is a unique integer number within this group.
2. Processes communicate with each other within a communicator. The default communicator for the group of processes started in the application is `MPI_COMM_WORLD`. There are two types of communicators: intra for communication within a group and intercommunicators, described in Section 4.1.15, for communication between groups.
3. The MPI specification allows for using threads within MPI processes with the level of support depending on a particular MPI implementation – see Section 4.1.14 for details. This support level refers to which threads of an MPI application are allowed to invoke MPI functions.

4.1.3 Initializing and finalizing usage of MPI

Each process of an MPI application needs to initialize the environment by calling:

```
int MPI_Init(int *argc,char ***argv)
```

with the following parameters:

- `argc` – pointer to a variable with the number of command line arguments,
- `argv` – pointer to an array of command line arguments

and finalize the code by calling:

```
int MPI_Finalize(void)
```

Similarly to other MPI functions, `MPI_SUCCESS` is returned in case there is no error.

Note that an alternative method of initializing the MPI environment is possible by calling `MPI_Init_thread` when threads are to be used within an MPI application, as discussed in Section 4.1.14.

4.1.4 Communication modes

In MPI, communication routines can be characterized in terms of working modes that define how many processes are involved, the way synchronization and actual execution of communication is performed.

Firstly, in MPI processes can send and receive messages in one of two modes:

1. Point-to-point – there are two processes involved in a communication routine. A process sends/receives a message to/from another process. In case of sending a message, a destination rank needs to be specified. In case of receiving a message, it is possible to specify a particular rank of the destination process or allow receiving from any process.

2. Collective/group – several processes are involved in communication and exchange of data. Processes in a communicator all take part in a group exchange of data or synchronization. Special hardware solutions might be used in implementation of such routines as compared to an implementation of such using point-to-point communication routines.

Independently from the above, communication routines can be characterized depending on how synchronization is performed, from the point of view of the calling process:

1. Blocking – the function blocks the execution of the process until the function has been executed (such as message received) or progressed until a certain step (such as copied to a final/additional buffer for a send operation such that the input buffer can be modified).

2. Non-blocking – the function returns immediately to the calling process and execution of the call may progress in the background. In this case, such a function can be thought of as a request for execution. There is a handle `MPI_Request` identified with the request which needs to be passed to additional functions for checking the status of a call or to terminate the call in a blocking way.

4.1.5 Basic point-to-point communication routines

The two basic communication functions include:

1. `MPI_Send` for sending a message to a destination process identified by a specific rank. `MPI_Send` is a blocking function. Specifically, following the MPI specification [6], the send operation will return when it is safe to modify data in the send buffer. An implementation may have copied data to a final destination or an intermediate buffer. The specification of the call is the following:

```
int MPI_Send(const void *buffer,int count,
MPI_Datatype datatype,int destination,int tag,
MPI_Comm comm)
```

with the following parameters:

- `buffer` – a pointer to a buffer that contains the message to be sent – the message can contain either a certain number of elements of a given simple data type or elements of several data types – see Section 4.1.7,
- `count` – the number of elements of the given data type,
- `datatype` – data type for elements used in the message, it is also possible to use a data type that encapsulates a complex structure containing several simple data types – see Section 4.1.7,
- `destination` – the rank of the destination process,
- `tag` – a label associated with the given message; usually tags are used in order to denote types of messages used in the application e.g. tag `DATA_DISTRIBUTION` for data distribution, tag `DATA_MERGING` for collection of data etc. (it is a good practice to use `#define` in order to define tags used throughout the application),
- `comm` – denotes the communicator within which communication is performed and the process identified.

2. `MPI_Recv` for a blocking receive of a message from a source process:

```
int MPI_Recv(void *buffer,int count,
MPI_Datatype datatype,int source,int tag,
MPI_Comm comm,MPI_Status *status)
```

with the following parameters:

- `buffer` – a pointer to a buffer for storing the incoming message, the buffer should have been allocated before,
- `count` – the number of elements of the given data type,
- `datatype` – data type for elements used in the message,
- `source` – the rank of a sender process, instead of a specific rank, `MPI_ANY_SOURCE` can be used for receiving from any process,
- `tag` – a label associated with the given message, instead of a specific tag, `MPI_ANY_TAG` can be used for receiving a message with any tag,
- `comm` – denotes the communicator within which communication is performed and process identified,
- `status` – contains information related to the message such as the rank of the sender and message tag.

4.1.6 Basic MPI collective communication routines

While communication among several processes can be implemented using point-to-point communication functions, it may be suboptimal and may introduce unnecessary complexity into the code. Several collective communication calls that are useful in a variety of applications are available. In MPI, in contrast to the point-to-point mode, all processes involved in a collective call invoke the same function. The meaning of some arguments may be different depending on the rank of a process. MPI functions from this group include:

1. Barrier – a group of processes is synchronized i.e. a process from the group is allowed to pass the barrier only after each of the other processes from the group has reached the barrier. The function is invoked by processes in the `comm` communicator:

   ```
   int MPI_Barrier(MPI_Comm comm)
   ```

2. Scatter – one or more processes distribute data to a group of processes. In the standard `MPI_Scatter`, one selected process sends chunks of data to a group of processes:

   ```
   int MPI_Scatter(const void *sendbuffer,int sendcount,
   MPI_Datatype senddatatype,void *receivebuffer,
   int receivecount,MPI_Datatype receivedatatype,
   int root,MPI_Comm comm)
   ```

   with the following parameters:

   - `sendbuffer` – a pointer to data that needs to be sent to processes, this value will be used only in the root process,
   - `sendcount` – how many elements should be sent to each process,
   - `senddatatype` – what data type is used for elements,
   - `receivebuffer` – a pointer to a preallocated space for the data to be received from the root process,
   - `receivecount` – the number of elements in `receivebuffer`,
   - `receivedatatype` – data type used for the data in `receivebuffer`,
   - `root` – the rank of the process distributing data,
   - `comm` – communicator for the group of processes involved in communication.

   Additionally, there are other versions of this basic scatter routine:

```
int MPI_Scatterv(const void *sendbuffer,
const int *sendcounts,const int *displacements,
MPI_Datatype senddatatype,void *receivebuf,
int receivecount,MPI_Datatype receivedatatype,
int root,MPI_Comm comm)
```

which allows sending data of various lengths to various processes. Additional parameters include:

- `sendcounts` – an array that specifies how many elements to send to each of the processes,
- `displacements` – an array with displacements for data to be sent to processes.

3. Gather – an operation opposite to scatter. In the basic `MPI_Gather` one selected root process gathers chunks of data from a group of processes. This particular call:

```
MPI_Gather(const void *sendbuffer,int sendcount,
MPI_Datatype senddatatype,void *receivebuffer,
int receivecount,MPI_Datatype receivedatatype,
int root,MPI_Comm comm)
```

has the following parameters:

- `sendbuffer` – a pointer to a buffer with the data to be sent,
- `sendcount` – the number of elements to be sent,
- `senddatatype` – the data type for elements to be sent,
- `receivebuffer` – a pointer to a buffer that will collect data gathered from the processes involved in communication,
- `receivecount` – the number of elements received from each process only valid for the root process,
- `receivedatatype` – data type for elements received in the receive buffer by the root process,
- `root` – the rank of the receiving process,
- `comm` – communicator for the group of processes involved in communication.

Similarly to `MPI_Scatter` and `MPI_Scatterv`, there is a version of `MPI_Gather` that allows receiving various data sizes from various processes:

```
int MPI_Gatherv(const void *sendbuffer,int sendcount,
MPI_Datatype senddatatype,void *receivebuffer,
const int *receivecounts,const int *displacements,
MPI_Datatype recvtype,int root,MPI_Comm comm)
```

with the following additional parameters:

- `receivecounts` – an array of integers that specify the numbers of elements received from the processes,
- `displacements` – an array of integers that specify displacements for data chunks to be received by the root process i.e. where to put the received data counting from the beginning of `receivebuffer`.

Additionally, the gather functions also include a version that allows all processes to receive data, not only the root process. This is possible by invoking function:

```
int MPI_Allgather(const void *sendbuffer,int sendcount,
MPI_Datatype senddatatype,void *receivebuffer,
int receivecount,MPI_Datatype receivedatatype,
MPI_Comm comm)
```

which allows each process to receive data chunks from each process in such a way that the data chunk from process with rank i is put into the i-th part of `receivebuffer`. Similarly to the aforementioned versions, `MPI_Allgatherv` allows receiving data chunks of various lengths:

```
int MPI_Allgatherv(const void *sendbuffer,int sendcount,
MPI_Datatype senddatatype,void *receivebuffer,
const int *receivecounts,const int *displacements,
MPI_Datatype receivetype,MPI_Comm comm)
```

4. Broadcast – in this case one process broadcasts data to all the processes in the communicator:

```
int MPI_Bcast(void *buffer,int count,
MPI_Datatype datatype,int root,MPI_Comm comm)
```

with the following parameters:

- `buffer` – data to be sent out by the root process,
- `count` – the number of elements to send from `buffer`,
- `datatype` – data type for the elements in `buffer`,

- root – the rank of the process that sends the data,
- comm – communicator for the group of processes involved in communication.

5. Reduce – in some cases it is preferable to perform operations on data that is gathered from processes involved in communication. MPI_Reduce allows communication and a given operation on data:

```
int MPI_Reduce(const void *sendbuffer,void *receivebuffer,
int count,MPI_Datatype datatype,MPI_Op op,
int root,MPI_Comm comm)
```

Specifically, each of the processes in the communicator provides a value or multiple values as input. Corresponding values from the processes are gathered and operation op is performed on them which results in an output value that is put into receivebuffer of the process with rank root. In case each process provides more than one value, operations are performed for corresponding values provided by processes. The meaning of parameters is as follows:

- sendbuffer – contains a value or values provided by each process involved in the operation,
- receivebuffer – space provided for results in the process with rank root,
- count – the number of values each process provides to the operation,
- datatype – data type of the elements for which values are provided,
- op – the operation performed on the values. MPI provides several predefined operations including:
    - MPI_MAX – maximum of values,
    - MPI_MIN – minimum of values,
    - MPI_PROD – product of values,
    - MPI_SUM – sum of values,
    - MPI_LXOR – a logical xor of values etc.

It should also be noted that it is possible to create a new operation using function:

```
int MPI_Op_create(MPI_User_function *user_function,
int iscommutative,MPI_Op *op)
```

6. All to all communication. In this case all processes send data to all other processes. This communication mode implemented by function `MPI_Alltoall` can be thought of as an extension of `MPI_Allgather`. In this particular case, each process sends various data to various receiving processes. Specifically [6] process with rank i sends the j-th chunk of data to process with rank j. The process with rank j receives this data chunk to the i-th chunk location. Details of this function are as follows:

```
int MPI_Alltoall(const void *sendbuffer,int sendcount,
MPI_Datatype senddatatype,void *receivingbuffer,
int receivecount,MPI_Datatype receivedatatype,
MPI_Comm comm)
```

   with the meaning of parameters such as for `MPI_Allgather`.

   Additionally, MPI specifies `MPI_Alltoallv` that allows sending data chunks of various sizes between processes. Otherwise, the way of placement of received data is analogous to `MPI_Alltoall`. However, specific placement of data chunks is provided with displacements:

```
int MPI_Alltoallv(const void *sendbuffer,
const int *sendcounts,const int *senddisplacements,
MPI_Datatype sendtype,void *receivebuffer,
const int *receivecounts,const int *receivedisplacements,
MPI_Datatype receivedatatype,MPI_Comm comm)
```

   Compared to the previous functions, additional parameters include:

   – `sendcounts` – an array of integers that specify how many elements should be sent to each process,

   – `senddisplacements` – an array that specifies displacements for data to be sent to particular processes i.e. `senddisplacements[i]` specifies the displacement for data to be sent to process i counting from pointer `sendbuffer`,

   – `receivedisplacements` – an array that specifies displacements for data to be received from particular processes i.e. `receivedisplacements[i]` specifies the displacement for data to be received from process i counting from pointer `receivebuffer`.

It should be noted that MPI implementations may optimize execution of such functions (as well as others) using hardware features of the system the application is run on.

4.1.7 Packing buffers and creating custom data types

In many cases, an implementation requires more complex messages than those that contain elements of only one data type. Such a need can be handled in MPI in the following ways:

1. Packing and unpacking elements of various data types into a buffer and passing a buffer to an `MPI_Send` type function and using the predefined `MPI_PACKED` type for sending the message.
2. Creating a custom data type that would already consider various data types, numbers of elements, strides etc. Such a type can be created using one of many MPI functions. It needs to be registered using function `MPI_Type_commit` and can subsequently be used in the standard MPI communication functions.

Before a buffer is created, it is necessary to obtain the size for the buffer so that it can contain all necessary elements. MPI provides function `MPI_Pack_size` that returns an upper bound on the size required to contain a specified number of elements of a given data type:

```
int MPI_Pack_size(int incount,MPI_Datatype datatype,
MPI_Comm comm,int *size)
```

with the following parameters:

- `incount` – the number of elements,
- `datatype` – datatype to be used when packing,
- `comm` – communicator to be used when packing,
- `size` – an upper bound on the size required to pack a message expressed in bytes.

Having obtained sizes for particular blocks of elements of various data types, the total size for a buffer can be computed by adding required sizes for the blocks. Subsequently, a buffer can be allocated using `malloc` and packed with blocks using `MPI_Pack` for each of the blocks:

```
int MPI_Pack(const void *inputbuffer,int incount,
MPI_Datatype datatype,void *outputbuffer,int outsize,
int *position,MPI_Comm comm)
```

with the following parameters:

- `inputbuffer` – pointer to the data to be packed into a buffer,
- `incount` – the number of elements to be packed,

- datatype – data type for the elements to be packed,
- outputbuffer – pointer to a buffer the data will be packed to,
- outsize – size of the output buffer in bytes,
- position – location in the output buffer where the data will be packed to, after data has been written to the location; after the data has been written position will point to the next location in the output buffer where new data can be stored,
- comm – communicator that will be used for the message.

Then the buffer can be sent using MPI_PACKED for the message type. On the receiver side, the message can be unpacked using MPI_Unpack:

```
int MPI_Unpack(const void *inputbuffer,int insize,
int *position,void *outputbuffer,int outcount,
MPI_Datatype datatype,MPI_Comm comm)
```

with the following parameters:

- inputbuffer – pointer to a buffer from which data should be unpacked,
- insize – the size of the input buffer in bytes,
- position – location in the inputbuffer from which data should be unpacked,
- outputbuffer – pointer to a buffer data will be unpacked to,
- outcount – the number of elements to be unpacked,
- datatype – data type for the elements to be unpacked,
- comm – communicator that was used for receiving this message.

Alternatively, an application may create its own custom data types that will be used throughout execution. This is especially desirable if a process requires to send a more complex data type than a simple structure in which blocks of elements are located one after another. For instance, sending a column of a matrix may require a single data type but skipping a certain number of bytes between element locations. MPI offers several functions for creating custom data types. Definition of a new data type should be followed by invocation of function MPI_Type_commit(MPI_Datatype *datatype) that registers the data type. Specific functions for data type definition include:

```
int MPI_Type_contiguous(int elemcount,
MPI_Datatype datatype,MPI_Datatype *newdatatype)
```

that creates a datatype that is composed of `elemcount` number of elements each of which is `datatype`. `newdatatype` is returned.

```
int MPI_Type_vector(int blockcount,int blocklength,
int stride,MPI_Datatype datatype,
MPI_Datatype *newdatatype)
```

creates a datatype that is composed of blocks of data type `datatype`, each of which `blocklength` long, spaced by stride (which denotes number of elements between the beginnings of successive blocks). `newdatatype` is returned. `MPI_Type_hvector` has the same signature as `MPI_Type_vector` with the exception of stride which is given in bytes and is of type `MPI_Aint`.

```
int MPI_Type_indexed(int blockcount,const int *blocklengths,
const int *displacements,MPI_Datatype datatype,
MPI_Datatype *newdatatype)
```

specifies that a new data type consists of `blockcount` blocks, each of which has length specified by array `blocklengths` and `displacements` which are expressed with the given number of elements of data type `datatype`. `newdatatype` is returned.

```
int MPI_Type_create_struct(int blockcount,
const int blocklengths[],const MPI_Aint displacements[],
const MPI_Datatype types[],MPI_Datatype *newtype)
```

is the most general function which allows definition of structures. Compared to `MPI_Type_indexed`, various types of elements in each block are allowed. `newdatatype` is returned. An example of using this function is presented in Section 5.2.1.

4.1.8 Receiving a message with wildcards

Quite often a process knows exactly the rank of the sender process. In this case, the rank can be used in an `MPI_Recv` function. However, in many cases it is not known in advance from which sender process the next message arrives. For instance, having sent input data to slave processes, a master will not know which slave will send back its result first. There can be several reasons for this fact:

1. slave processes may run on various CPUs,
2. some nodes where slaves are running may become overloaded by other processes,
3. data packets take various amounts of time to be updated.

In such cases, a process might want to receive a message from any process, regardless of which sends a message first. This can be accomplished by calling function `MPI_Recv` with `MPI_ANY_SOURCE` specified as the rank of the sender. Obviously, the receiving process may need to find out the actual rank of the sender in order to communicate with it next. This can be achieved as follows:

```
MPI_Status status;

MPI_Recv(buffer,count,datatype,MPI_ANY_SOURCE,tag,
communicator,&status);
// find out the real sender rank
int senderrank=status.MPI_SOURCE;
```

Similarly, a process may wait for a message with a given tag or for a message with any tag. It can be achieved in an analogous way using `MPI_ANY_TAG` for the tag in `MPI_Recv`. The actual tag of the received message can be read as follows after `MPI_Recv`:

```
int tag=status.MPI_TAG;
```

4.1.9 Receiving a message with unknown data size

Many applications send and receive messages of predefined formats i.e. the number and even sizes of blocks of elements of various data types. However, in some cases a receiver does not know the size of an incoming message. This needs to be handled because if the whole message needs to be received then a buffer of sufficient size has to be allocated in advance. MPI allows handling such a scenario in the following way:

1. Find out about the incoming message without actually receiving it i.e. check if such a message is available. It can be done using blocking `MPI_Probe` or non-blocking `MPI_Iprobe` functions:

   ```
   int MPI_Probe(int source,int tag,
   MPI_Comm comm,MPI_Status *status)
   ```

   with the following parameters:

   - `source` – rank of the sender process, `MPI_ANY_SOURCE` can also be used as discussed in Section 4.1.8,
   - `tag` – a label associated with the given message. `MPI_ANY_TAG` can also be used as discussed in Section 4.1.8,
   - `comm` – denotes the communicator within which communication is performed and process identified,

- `status` – contains information related to the message.

```
int MPI_Iprobe(int source,int tag,
MPI_Comm comm,int *flag,MPI_Status *status)
```

with the following parameters:

- `source` – rank of the sender process,
- `tag` – a label associated with the given message,
- `comm` – denotes the communicator within which communication is performed and process identified,
- `flag` – true i.e. other than 0 if a message with the specified (previous) parameters is available,
- `status` – contains information related to the message.

2. Find out the size of the incoming message by invoking:

```
int MPI_Get_count(const MPI_Status *status,
MPI_Datatype datatype,int *count)
```

with the following parameters:

- `status` – information about the message returned by `MPI_Probe` or `MPI_Iprobe`,
- `datatype` – a data type that will be used when allocating a receive buffer,
- `count` – the number of elements in the message.

3. Allocate a buffer of size indicated by `MPI_Get_count`. This can be done with the `malloc` function.

4. Receive the message using one of `MPI_Recv` family functions into the preallocated buffer.

4.1.10 Various send modes

MPI distinguishes several send modes and corresponding functions. These include:

- `MPI_Send(...)` – a basic blocking send function. After the function has returned, the calling process is allowed to modify the send buffer. MPI can buffer the message which may cause the function to return fast. Alternatively, this call may even block until the moment the sent message is buffered on the receiver side.

- `MPI_Bsend(...)` – a buffered send. In this case when there has been no receive call on the addressee side, the MPI runtime will need to buffer the message or report an error if not enough space is available.

- `MPI_Ssend(...)` – a synchronous send. In addition to the standard send semantics, `MPI_Ssend(...)` will complete only after a matching receive has been called on the addressee side and the receiver has started the process of receiving the message. The sender process can reuse the buffer after the function returned.

- `MPI_Rsend(...)` – a ready send mode. In this case, it is assumed (and it is up to the programmer to ensure such a condition) that a receiver has already called a matching receive. This potentially allows to perform communication faster as the destination buffer is already known. If no matching receive has been invoked then an error occurs.

4.1.11 Non-blocking communication

MPI specifies functions that allow start of send and receive operations and performing these in the background while allowing the calling thread to continue. This is especially useful for implementing overlapping communication and computations described in Section 6.1.1. The basic `MPI_Send` and `MPI_Recv` have the following non-blocking variants:

```
int MPI_Isend(const void *buffer,int count,
MPI_Datatype datatype,int destination,int tag,
MPI_Comm comm,MPI_Request *request)
```

with the following parameters:

- `buffer` – a pointer to a buffer that contains the message to be sent,
- `count` – the number of elements of the given data type,
- `datatype` – data type for elements used in the message,
- `destination` – the rank of the destination process,
- `tag` – a label associated with the given message,
- `comm` – denotes the communicator within which communication is performed and process identified,
- `request` – a handler associated with the call that can be used for completion of the call later.

```
int MPI_Irecv(void *buffer,int count,
MPI_Datatype datatype,int source,
int tag,MPI_Comm comm,MPI_Request *request)
```

with the following parameters:

- `buffer` – a pointer to a buffer for storing the incoming message, the buffer should have been allocated before,
- `count` – the number of elements of the given data type,
- `datatype` – data type for elements used in the message,
- `source` – the rank of a sender process,
- `tag` – a label associated with the given message,
- `comm` – denotes the communicator within which communication is performed and process identified,
- `request` – a handler associated with the call that can be used for completion of the call later.

Starting `MPI_Isend` or `MPI_Irecv` means that actual communication may start from that point on and progress in the background. Consequently, the buffer passed in `MPI_Isend` and the buffer in `MPI_Irecv` should not be written to or read from until the corresponding non-blocking operation has completed.

Functions `MPI_Wait` or `MPI_Test` can be invoked to complete `MPI_Irecv` or `MPI_Isend`. As indicated in [6] it does not mean that the message was received. Actually, it means that the aforementioned buffers can now be safely modified in case of `MPI_Isend` or read from in case of `MPI_Irecv`. `MPI_Wait` is a blocking call that returns after the non-blocking call associated with request has completed:

```
int MPI_Wait(MPI_Request *request,MPI_Status *status)
```

with the following parameters:

- `request` – a request associated with a previous call,
- `status` – a status object associated with the call.

It should be noted that after the function has completed, the request is either set to inactive or `MPI_REQUEST_NULL`. If `MPI_Wait` is invoked with `MPI_REQUEST_NULL` then it has no effect.

`MPI_Test`, on the other hand, is a non-blocking call. This means that `MPI_Test` returns a flag whether a previous message associated with a given request has completed. The function call is as follows:

```
int MPI_Test(MPI_Request *request,
int *flag, MPI_Status *status)
```

with the following parameters:

- `request` – a request associated with a previous call,
- `flag` – indicates whether a function associated with the request has completed,
- `status` – a status object associated with the call.

If the flag is set to true then, similarly to `MPI_Wait`, the request is either set to inactive or `MPI_REQUEST_NULL` while `status` will contain information on the call.

It should be noted that MPI actually provides a whole family of `MPI_Wait` and `MPI_Test` functions that allow passing more than just one request as input. These can be very useful in more complex communication patterns that involve more processes. These functions include:

```
int MPI_Waitany(int count,MPI_Request arraywithrequests[],
int *index, MPI_Status *status)
```

that allows blocking waiting for completion of any of the calls associated with requests given as input in `arraywithrequests`. A value of `index` will denote the one that completed. It should be noted that using `MPI_Waitany()` when handling communication with many processes might lead to some being served more frequently than others.

Another useful function is blocking `MPI_Waitall` that allows waiting for completion of all non-blocking calls for requests were given as input. This function can include requests for both send and receive calls. The syntax is as follows:

```
int MPI_Waitall(int count,MPI_Request arraywithrequests[],
MPI_Status arraywithstatuses[])
```

4.1.12 One-sided MPI API

The idea of one-sided communication API is to implement remote memory access and potentially increase concurrency over the traditional message passing communication paradigm [108]. In the one-sided model, a process may expose a so-called window in memory so that other processes can read or write from/to such a window using e.g. get/put like operations. Synchronization is separated from communication. MPI distinguishes two modes regarding synchronization, assuming origin and target processes are distinguished:

- passive target communication – only an origin process is directly involved in communication,
- active target communication – two sides are engaged in communication.

MPI allows the creation of windows using several functions. For instance,

```
int MPI_Win_create(void *address,MPI_Aint windowsize,
int displacementunit,MPI_Info info,
MPI_Comm communicator,MPI_Win *window)
```

allows each process in `communicator` (the call is collective) to expose its own window in memory for one-sided (remote memory access) operations. The function has the following parameters:

- `address` – pointer to a memory window to be exposed by each process,
- `windowsize` – size of the window in bytes,
- `displacementunit` – displacement size expressed in bytes,
- `info` – information object,
- `communicator` – communicator shared by the processes,
- `window` – returned window.

Other window definition functions include:

```
int MPI_Win_allocate(MPI_Aint windowsize,int displacementunit,
MPI_Info info,MPI_Comm communicator,
void *address,MPI_Win *window)
```

which is similar to `MPI_Win_create` but it allocates memory in each process; consequently the function returns a pointer and a window in each process.

```
int MPI_Win_allocate_shared(MPI_Aint size,
int displacementunit,MPI_Info info,
MPI_Comm communicator,void *address,MPI_Win *window)
```

which is similar to `MPI_Win_allocate` but all processes may access the memory using write/read operations. `address` can be used on the invoking process. By default, a contiguous memory across process ranks is allocated with at least size bytes in each rank. Function:

```
int MPI_Win_shared_query(MPI_Win window,int rank,
MPI_Aint *size,int *displacementunit,void *address)
```

allows the determination of details of a window previously created using MPI_Win_allocate_shared, specifically the size of the region for rank as well as the address of a shared region for the given rank. According to the MPI specification [6], the application should make sure that processes involved in the call can create a shared memory region.

```
int MPI_Win_create_dynamic(MPI_Info info,
MPI_Comm communicator,MPI_Win *window)
```

allows the creation of a window without attached memory. Memory can be then attached using function:

```
int MPI_Win_attach(MPI_Win window,void *address,
MPI_Aint memorysize)
```

with the following parameters:

- window – the window considered,
- address – pointer to the memory region to be attached,
- memorysize – size of the memory region to be attached, specified in bytes.

Subsequently, when needed, a memory region can be detached using function:

```
int MPI_Win_detach(MPI_Win window,const void *address)
```

where address specifies the region to be detached.

Basic functions allowing data manipulation within windows include: MPI_Put, MPI_Rput for writing data to a window. Specifically,

```
int MPI_Put(const void *originaddress,int originelemcount,
MPI_Datatype originelemdatatype,int targetrank,
MPI_Aint targetdisplacement,int targetelemcount,
MPI_Datatype targetelemdatatype,MPI_Win window)
```

allows the specification of a send like operation to a target window's memory. One process, on the origin side, specifies all parameters referring both to the origin and the target sides:

- originaddress – address of the buffer on the origin side,
- originelemcount – the number of elements on the origin side,
- originelemdatatype – data type of the elements on the origin side,
- targetrank – rank of the target process,

- `targetdisplacement` – displacement on the target side (counting from the beginning of the window),
- `targetelemcount` – the number of elements for the target side,
- `targetelemdatatype` – data type of the elements on the target side,
- `window` – the window used for the one-sided write operation.

On the other hand,

```
int MPI_Rput(const void *originaddress,int originelemcount,
MPI_Datatype originelemdatatype,int targetrank,
MPI_Aint targetdisplacement,int targetelemcount,
MPI_Datatype targetelemdatatype,MPI_Win window,
MPI_Request *request)
```

is a non-blocking version of `MPI_Put`. It should be noted that local completion of the former with the standard functions only ensures that the origin can modify its buffers. Completion on the target side can be performed e.g. with

```
int MPI_Win_flush(int rank, MPI_Win window)
```

which finalizes one-sided calls on both sides. `MPI_Get`, `MPI_Rget` allow reading data from a window. The syntax of `MPI_Get` is as follows:

```
int MPI_Get(void *originaddress,int originelemcount,
MPI_Datatype originelemdatatype,int targetrank,
MPI_Aint targetdisplacement,int targetelemcount,
MPI_Datatype targetelemdatatype, MPI_Win window)
```

where `origin*` variables concern the receive buffer on the origin side while `target*` variables for the memory data is received from. Similarly to `MPI_Rput`,

```
int MPI_Rget(void *originaddress,int originelemcount,
MPI_Datatype originelemdatatype,int targetrank,
MPI_Aint targetdisplacement,int targetelemcount,
MPI_Datatype targetelemdatatype,
MPI_Win window,MPI_Request *request)
```

returns a request that can be completed with standard completion functions indicating presence of data at `originaddress`.

Following the aforementioned synchronization modes, the following functions can be used:

- Passive target:

```
int MPI_Win_lock(int lockmode,int rank,
int assert,MPI_Win window)
```

with the following parameters:

- `lockmode` – can be either `MPI_LOCK_SHARED` in which case many processes can access the window (useful for `MPI_Get`) or `MPI_LOCK_EXCLUSIVE`,
- `rank` – rank subject to the operation,
- `assert` – used to define special conditions that can be used for optimization, passing 0 would mean that no conditions are implied [6],
- `window` – concerned window.

- Active target:
  - With collective fence operations. In this case, processes may access memories of other processes associated with a window. This synchronization mode can be used e.g. for geometric SPMD applications such as those discussed in Section 3.3 for the communication phase that follows a computation phase in iterations of an algorithm. The syntax of the fence operation is as follows:

    ```
    int MPI_Win_fence(int assert,MPI_Win window)
    ```

    with the following parameters:

    - `assert` – indicates special conditions for optimization of the operation. As indicated above, 0 implies no conditions,
    - `window` – the window which the operation concerns.
  - Synchronization in which a target process exposes its memory by calls to post and wait operations while an origin process accesses data in between calls of start and complete operations. These can be used for pair-wise synchronization. Particular calls are as follows:

    ```
    int MPI_Win_post(MPI_Group originprocessgroup,
    int assert,MPI_Win window)
    ```

    with the following parameters:

    - `originprocessgroup` – a group of processes on the origin side,
    - `assert` – conditions that may be used to optimize the call, 0 implies no conditions,
    - `window` – the window which the operation concerns.

```
int MPI_Win_wait(MPI_Win window)
```

where `window` is the window which the operation concerns.

```
int MPI_Win_start(MPI_Group targetprocessgroup,
int assert,MPI_Win window)
```

with the following parameters:

- `targetprocessgroup` – indicates a group of processes on the target side,
- `assert` – conditions that may be used to optimize the call, 0 implies no conditions,
- `window` – the window which the operation concerns.

```
int MPI_Win_complete(MPI_Win window)
```

where `window` is the window which the operation concerns.

4.1.13 A sample MPI application

Let us assume the problem of computing the following series:

$$\ln(x) = \frac{x-1}{x} + \frac{(\frac{x-1}{x})^2}{2} + \frac{(\frac{x-1}{x})^3}{3} + \ldots \quad (4.1)$$

for $x > \frac{1}{2}$. Such a problem can be parallelized by assignment of an equal number of elements of the sum to each process of an MPI application. It should also be noted that each successive element of the sum seems to involve more computations since the power increases. One possible solution, that can be adopted, is as follows:

1. Process with rank i computes elements i, $i+p$, $i+2p$, $i+3p$, where p is the number of processes started in the application.
2. It can be noted that each process can precompute the value of x^p that can be used for computing the value of the next element.

Alternatively, each process may deal with a range of consecutive elements of the series. If values of successive elements are considerably smaller and smaller, this might be a preferred approach in terms of accuracy of values added within a process. The number of elements required to reach a good accuracy might be different for various values of argument x when computing $\ln(x)$. Such an MPI program for parallel computing of $\ln(x)$ is included in file `lnx_MPI.c`. An important part of such an implementation is shown in Listing 4.1.

Listing 4.1 Sample parallel MPI program for computing $\ln(x)$

```
int i;
double x;
int myrank,nproc;
long maxelemcount=10000000;
double partialsum=0; // partial sum computed by each process
double mult_coeff; // by how much multiply in every iteration
double sum=0; // final value meaningful for process with rank
  0
double prev; // temporary variable
long power; // acts as a counter
long count; // the number of elements

// initialize MPI with pointers to main() arguments
MPI_Init(&argc,&argv);

// obtain my rank
MPI_Comm_rank(MPI_COMM_WORLD,&myrank);
// and the number of processes
MPI_Comm_size(MPI_COMM_WORLD,&nproc);

(...)
// read the argument
x=...

// read the second argument
maxelemcount=...

mult_coeff=(x-1)/x;
count=maxelemcount/nproc;
// now compute my own partial sum in a loop
power=myrank*count+1;
prev=pow(((x-1)/x),power);
for(;count>0;power++,count--) {
  partialsum+=prev/power;
  prev*=mult_coeff;
}

// now each process should have its own partial sum ready
// add the values to the process with rank 0
MPI_Reduce(&partialsum,&sum,1,MPI_DOUBLE,MPI_SUM,0,
  MPI_COMM_WORLD);

if (!myrank) // print out the result
  printf("\nResult is %f\n",sum);
```

```
MPI_Finalize();
```

For such an application it should be noted that:

1. Each process can determine the input data it should process on its own. No data distribution phase was needed which is often necessary if e.g. only one process has access to input data or one process generates input data.

2. Processes can perform operations independently and partial results can be merged into a final value at the end of computations. Such an application is often referred to as embarrassingly parallel. In many problems, communication and synchronization among processes are much more complex, as described in Chapter 3.

3. In this specific example computations are balanced among processes. In many practical problems, as described in Chapter 3, load balancing will be a real challenge and may even need to be performed at runtime.

Compilation of the application and running can be performed as follows:

```
mpicc <flags> lnx_MPI.c -lm
mpirun -np 16 ./a.out 300236771.4223 4000000000
```

Results for the application run with parameters 300236771.4223 20000000000 on a workstation with 2 x Intel Xeon E5-2620v4 and 128 GB RAM are shown in Figure 4.2. Results are best values out of three runs for each test. OpenMPI 2.0.1 was used.

It should be noted that MPI is typically used for running parallel applications on a cluster of nodes to enable efficient communication between nodes but can also allow communication between processes running on one node.

4.1.14 Multithreading in MPI

Apart from an MPI application being composed of many processes, the MPI specification also allows threads within processes. Specifically, various support levels for multithreading, i.e. allowing calling MPI functions from threads, are distinguished:

- `MPI_THREAD_SINGLE` – no support is offered for multithreading from an MPI implementation,
- `MPI_THREAD_FUNNELED` – in this case there is a restriction allowing only the thread that initialized MPI to invoke MPI functions,
- `MPI_THREAD_SERIALIZED` – only one thread at a time is allowed to call an MPI function,

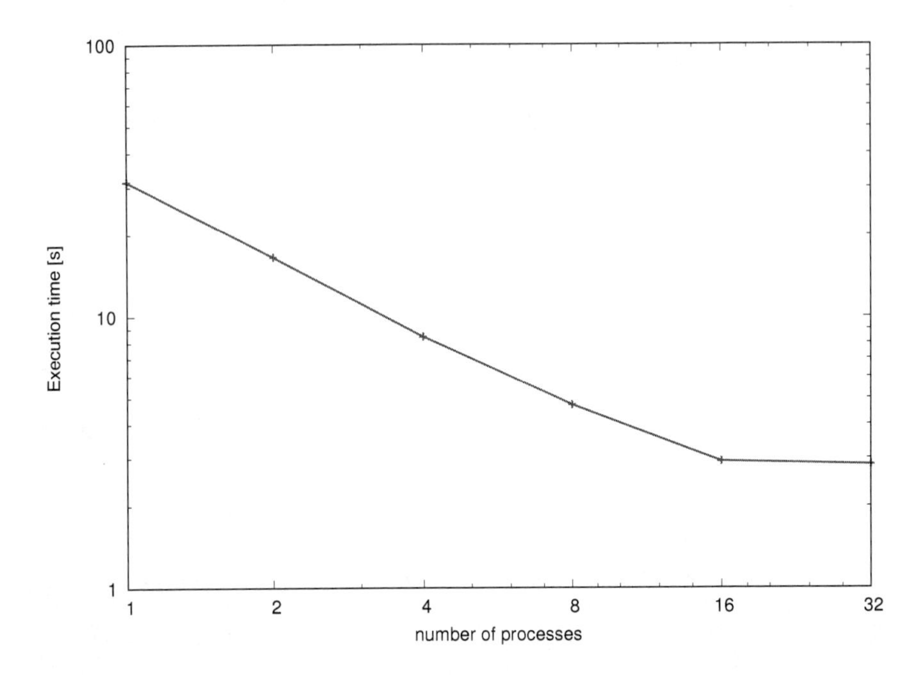

FIGURE 4.2 Execution time of the testbed MPI application run on a workstation with 2 x Intel Xeon E5-2620v4, 128 GB RAM, 16 physical cores, 32 logical processors, parameters: 300236771.4223 20000000000

– `MPI_THREAD_MULTIPLE` – in this case there are no restrictions – any thread within a process can call MPI functions.

In order to use many threads, the MPI environment should be initialized using function `MPI_Init_thread` instead of `MPI_Init`. The former has the following syntax:

```
int MPI_Init_thread(int *argc,char ***argv,
int required,int *provided)
```

and parameters:

– `argc` – pointer to a variable with the number of command line arguments,

– `argv` – pointer to an array of command line arguments,

– `required` – desired support level of multithreading (expressed using the aforementioned levels),

– `provided` – returned level supported by the given MPI implementation.

Examples of MPI applications using multithreading are presented in Sections 4.7.1 and 4.7.2.

4.1.15 Dynamic creation of processes in MPI

Previously dynamic process creation was one of the advantages of a very flexible system called Parallel Virtual Machine [73] that also enabled adding or removing hosts to/from the virtual machine.

MPI also allows dynamic creation of new processes while an application is already running. Furthermore, it allows establishing communication between the already existing and new processes. From the point of view of the application, processes are started and the application does not interact with the operating system and underlying process management such as queuing systems.

MPI specifies two main functions for dynamic spawning of processes:

```
int MPI_Comm_spawn(const char *commandtospawn,char *argv[],
int maxproccount,MPI_Info info,
int root,MPI_Comm communicator,
MPI_Comm *intercommunicator,int arrayoferrcodes[])
```

with the following parameters:

- `commandtospawn` – name of the executable that should be spawned,
- `argv` – command line arguments,
- `maxproccount` – the maximum number of processes that should be spawned,
- `info` – information that can include hints as to where to start processes, an example is shown in Section 5.3.3.2 – specifically, an `MPI_Info` object is created with function `MPI_Info_create` that will be subsequently passed to `MPI_Comm_spawn`; before that function `MPI_Info_set(info,"host","<hostname>")` can be used to set the name of the target host to `<hostname>`,
- `root` – rank of the process in which values of the previous arguments should be set,
- `communicator` – communicator that involves the spawning processes,
- `intercommunicator` – intercommunicator that will allow communication between the group of spawning processes and the group of spawned processes,
- `arrayoferrcodes` – array of codes of errors concerning particular processes.

The other function allows for specifying various executables to be used while spawning:

```
int MPI_Comm_spawn_multiple(int commandcount,
char *commandstospawn[],char **arrayofargvs[],
const int maxprocstospawn[],const MPI_Info arrayofinfos[],
int root,MPI_Comm communicator,
MPI_Comm *intercommunicator,int arrayoferrcodes[])
```

with additional/different parameters compared to `MPI_Comm_spawn`:

- `commandcount` – the number of commands in the next argument,
- `commandstospawn` – array of executables to be run,
- `arrayofargvs` – array of command line arguments,
- `maxprocstospawn` – array that specifies the maximum number of processes to be spawned for each executable,
- `arrayofinfos` – array of information objects that can include hints as where to start processes.

Error codes would be different from `MPI_SUCCESS` in case of error. It should be noted that the functions are collective over `communicator`. As an example, a group of processes may launch a new process or processes in which case e.g. `MPI_COMM_WORLD` may be used. Alternatively, this can be done from the point of view of a single process using `MPI_COMM_SELF`.

After a new process or processes (so-called children) have been spawned, the original group of processes and children may communicate using MPI functions such as `MPI_Send` or `MPI_Recv` using `intercommunicator`. Children may be addressed using ranks starting from 0 when using `intercommunicator`.

On the other hand, children would fetch a parent communicator using function:

```
int MPI_Comm_get_parent(MPI_Comm *parentcommunicator)
```

which returns the parent communicator. A process has been been spawned by another process if `parentcommunicator!=MPI_COMM_NULL`. For instance, if a single process started children, it could be contacted with using rank 0 and `parentcommunicator`.

When using arguments to be passed to spawned processes, `argv` contains just arguments to be passed, without the name of the binary. Additionally, the info argument may be used to pass information when a process should be started. As an example, the following sequence sets a host name for launch:

```
MPI_Info info;

MPI_Info_create(&info);
MPI_Info_set(info,"host","<hostname>");
```

```
MPI_Comm_spawn("binary",argv,1,info,0,MPI_COMM_SELF,
&intercommunicator,&errorcode);
```

An example that uses `MPI_Comm_spawn` for dynamic launching of processes and load balancing is presented in Section 5.3.3.2.

Function:

```
int MPI_Comm_disconnect(MPI_Comm *communicator)
```

can be used for collective internal completion of pending communication concerning `communicator` and frees the communicator object.

4.1.16 Parallel MPI I/O

Parallel MPI I/O specification [6] allows processes of an MPI application to open a file collectively and operate on data in the file using a variety of read-/write operations in several modes. A file is opened with a collective call to function:

```
int MPI_File_open(MPI_Comm communicator,const char *filename,
int amode,MPI_Info info,MPI_File *filehandle)
```

with the following parameters:

- `communicator` – communicator associated with a group of processes,
- `filename` – name of a file to open,
- `amode` – access mode; typical access modes can be specified with the following constants, among others: `MPI_MODE_RDWR` – for read and write access, `MPI_MODE_RDONLY` – open in read only mode, `MPI_MODE_WRONLY` – open for writes only, `MPI_MODE_CREATE` – create a file in case it is not present,
- `info` – can be used to indicate file system information that can allow optimization; otherwise, a programmer can provide `MPI_INFO_NULL`,
- `filehandle` – file handle that is associated with the file and used in successive calls.

It should be noted that the processes must refer to the same file with the same access mode. A file should be closed with a call to function:

```
int MPI_File_close(MPI_File *filehandle)
```

where `filehandle` is associated with a file.

The specification distinguishes several functions grouped with respect to:

1. File pointer(s) in a file. The following modes are possible in terms of referencing location in a file:
   - a given offset in the file,
   - a process individual file pointer,
   - a shared file pointer.
2. How many processes are involved in a call. Similarly to communication functions, two groups are distinguished:
   - individual calls from a process,
   - collective calls in which processes read or write parts of data in parallel.

4.2 OPENMP

4.2.1 Programming model and application structure

OpenMP is an API [126, 127] that allows easy extension of sequential applications such as written in C or Fortran so that execution of selected regions within the code is performed in parallel. In essence, an application can be extended with the following constructs:

1. directives that specify regions to be parallelized along with additional information including:
   - scoping of variables,
   - initialization of selected variables,
   - details on how iterations of loops are assigned to threads,
2. library functions that allow e.g. fetching the number of threads executing a parallel region or the id of a calling thread.

Additionally, environment variables can be set to impact running of a parallel application. Essentially, the structure of an a C+OpenMP application is shown in Listing 4.2, with OpenMP directives that specify following regions executed in parallel. Consequently, the OpenMP programming model distinguishes regions executed in parallel separated by parts of the code executed sequentially – such as for data aggregation and subsequent partitioning for parallel processing.

Listing 4.2 Basic structure of an OpenMP application

```
#include <omp.h>
#include <stdio.h>

int main() {
```

```
  // a serial region
  printf("\nThis is executed by one thread only!");
  // serial processing here


  // a parallel region with optional parameters [...]
#pragma omp parallel [...]
  {
    printf("\nThis code is potentially executed by several
    threads");
    // parallel processing here
  }

  // a serial region
  printf("\nThis code is executed by one thread only");
  // serial processing here

  // a parallel region with optional parameters [...]
#pragma omp parallel [...]
  {
    printf("\nThis code is potentially executed by several
    threads");
    // parallel processing here
  }

  // a serial region
  printf("\nThis code is executed by one thread only");
  // serial processing here

}
```

Directives may have several attributes and arguments that specify how parallelization within particular regions is performed. As an example, the pseudocode shown in Listing 4.3 distinguishes two parallel regions. 32 threads are requested (`num_threads(32)`) for execution of the first one. In the second, the number of threads will be selected by the runtime system unless specified by an environment variable as discussed in Section 4.2.3. The second parallel region specifies scoping of variables within the region – `variable0` and `variable1` are variables private to each thread while `variable2` is shared and will require proper synchronization when written to by threads.

Listing 4.3 Basic OpenMP structure with numbers of threads and function calls

```
#include <omp.h>
```

```
#include <stdio.h>

int main() {

  // a serial region
  printf("\nThis is executed by one thread only!");
  // serial processing here


  // a parallel region
  // each thread can fetch its own id unique
  // among the threads and the number
  // of threads executing the region (set to 32)
#pragma omp parallel num_threads(32)
  {
    printf("\nHello from thread %d, nthreads %d\n",
    omp_get_thread_num(), omp_get_num_threads());

    // parallel processing here

  }

  // a serial region
  printf("\nThis code is executed by one thread only");
  // serial processing here

  // a parallel region
  // the following specifies a parallel region with scoping of
    variables
  // in this case the number of threads is not set a'priori
#pragma omp parallel private(variable0, variable1) shared(
    variable2)
  {
    // parallel processing here
  }

  // this is executed by one thread only

}
```

4.2.2 Commonly used directives and functions

There are a few key directives that are typically used for parallelization of blocks in the code. These include:

– parallelization of a block by starting a team of threads in which the running thread becomes a master with thread number equal to 0. Each of the threads will execute the block.

```
#pragma omp parallel
<block>
```

Each of the threads can fetch the number of threads active in the team by calling function:

```
int omp_get_num_threads()
```

as well as its own id within the team by calling

```
int omp_get_thread_num()
```

These values would be typically used by the threads to assign parts of work for parallel execution of disjoint data sets. It should be noted that at the end of the block threads would synchronize by default. It is important to note that #pragma omp parallel can be extended with clauses that specify data scoping as well as potential reduction operations, the number of threads executing the block and a condition. These can be done by inserting one or more of the following:

- private(varlist) where varlist is a list of variables that will be private to each thread in the team. Variables are separated by a comma.
- shared(varlist) where varlist is a list of variables that will be shared among threads in the team and consequently access would need to be synchronized. Variables are separated by a comma.
- firstprivate(varlist) where varlist is a list of variables that will be private to each thread in the team and initial values will be the same as of these variables before execution of the block in parallel. Variables are separated by a comma.
- reduction(operator:varlist) performs a collective operation on the variables in the list. Specifically, it is assumed that there will be private copies of each variable on the list initialized with a value depending on the operator identifier. At the end of the block, the specified operation will be performed using these private copies for each variable and the result will be stored in a global variable. In particular, frequently used operators for forms var=var <operator> expression include:
  – + – addition,
  – * – multiplication,

- `-` – subtraction,
- `&` – bitwise and,
- `|` – bitwise inclusive or,
- `^` – bitwise exclusive or,
- `&&` – logical and,
- `||` – logical or.

As an example, parallel sum can be computed using the code included in file `openmp-reduction-sum.c`. An important part of such an implementation is shown in Listing 4.4.

Listing 4.4 Basic OpenMP implementation of reduction with sum

```
  long int result=0;
  int counter,countermax=128;

#pragma omp parallel for shared(countermax) reduction
    (+:result)
  for(counter=0;counter<countermax;counter++)
    result+=compute(counter);

  printf("Final result=%ld\n",result);

  (...)
```

compiled and run as follows:

```
gcc <flags> -fopenmp openmp-reduction-sum.c -o omprsum
./omprsum
```

The loop iteration variable is automatically assumed private.

The maximum of computed values in iterations can be computed using code included in file `openmp-reduction-max.c` compiled and run as follows:

```
gcc <flags> -fopenmp openmp-reduction-max.c -o omprmax
./omprmax
```

The reduction code with OpenMP is shown in Listing 4.5.

Listing 4.5 Basic OpenMP implementation of reduction with maximum value

```
 long int result=0;
```

```
int id;

(...)
#pragma omp parallel private(id) reduction(max:result)
   num_threads(32)
{
  id=omp_get_thread_num();
  result=compute(id);
}
printf("Final result=%ld\n",result);

(...)
```

- `if (logical_expression)` – allows the specification of a condition so that only if the condition is met (`logical_expression` is true) for threads to be created; otherwise a single master thread will execute the code. It allows specification that parallelization should not be performed if e.g. data size is too small.
- `num_threads(expression)` – allows specification of the number of threads to be used within the block.

– parallelization of frequently used `for` loops. This can be done using:

```
#pragma omp for
<forloop>
```

Essentially, parallelization is performed by execution of iterations by various threads assuming a parallel region was created before i.e.:

```
#pragma omp parallel [...]
{
#pragma omp for
<forloop>
}
```

`#pragma omp for` can be extended with the following clauses:

- `private(varlist)` – as above.
- `firstprivate(varlist)` – as above.
- `lastprivate(varlist)` where varlist is a list of variables that will be private to each thread in the team and the final global value of the variable will be from the last iteration of the loop.
- `ordered` – if specified then iterations of the loop will be as in a sequential code.

- `nowait` – if specified then the threads would not need to synchronize at the end of the loop.
- `schedule(assignment_type[,chunk_size])` defines how iterations of the loop are assigned to the threads of a team. The following types can be specified:
  - `static` – if a chunk size is specified then each thread is assigned data chunks of size `chunk_size` statically; otherwise successive iterations are divided into chunks such that are distributed as equally as possible among the threads.
  - `dynamic` – chunks of the given size (or 1 if not specified) are distributed among threads in a dynamic fashion; i.e. after a thread has finished processing of a data chunk, it will fetch another.
  - `guided` – chunks are variable in size and are proportional to the number of iterations remaining divided by the number of threads. The given value denotes a minimum chunk size.
  - `auto` – the runtime system or the compiler decides about suitable mode.
  - `runtime` – the `OMP_SCHEDULE` environment variable will specify a mode (see Section 4.2.5).
- parallelization through specifications of blocks to be executed by threads of a team. This can be accomplished through the `sections` directive. The syntax is as follows:

```
#pragma omp sections [...] [clause ...]
{
  #pragma omp section
<block>
  .
  .
  .
  #pragma omp section
<block>

}
```

  Particular sections will be executed by threads in a team. The `sections` directive accepts aforementioned clauses such as: `private(...)`, `firstprivate(...)`, `lastprivate(...)`, `nowait`, `reduction(...)`.

4.2.3 The number of threads in a parallel region

In an OpenMP application, the number of threads can be specified in a few ways, the latter settings specified at more fine grained levels and overriding previously set values:

1. setting the OMP_NUM_THREADS environment variable, e.g. in the following way in Linux:

   ```
   export OMP_NUM_THREADS=32
   ./application
   ```

2. using the omp_set_num_threads() library function,
3. using the num_threads(<number of threads>) clause in directives such as #pragma omp parallel.

If not specified, an implementation will use a number such as the number of processors on a node.

4.2.4 Synchronization of threads within a parallel region and single thread execution

OpenMP features several constructs that allow synchronization of threads within a parallel region. Care should be taken which of these is used for what purpose as performance differences may be seen:

– #pragma omp critical – precedes a block of code that will be executed by one thread at a time. It specifies a critical section. There can be two types of critical sections:

  - unnamed – if only

    ```
    #pragma omp critical
    <block of code>
    ```

    is used then such a section is called unnamed. This means that all threads will synchronize on entry to the code. Secondly, if there are more unnamed sections, these would be regarded as one.
  - named – in this case it is possible to specify distinct critical regions in an application each of which is identified with a name specified as follows:

    ```
    #pragma omp critical <name>
    <block of code>
    ```

– `#pragma omp atomic` – precedes a memory update instruction (such as incrementing a value) that will be executed atomically. It would usually be faster then a critical section.

In some cases, if a large number of threads is run such as on an Intel Xeon Phi (co)processors, it may be beneficial to partition all threads into groups that would synchronize within groups. That can minimize synchronization overhead at the cost of load balancing flexibility as load needs to be assigned to such groups prior to computations that involve synchronization. This is shown in more detail in Section 6.3.2.

There are directives each of which precedes a block that is executed by one thread:

– `#pragma omp master` – specifies a block of code to be executed by the master thread,

– `#pragma omp single` – specifies a block of code to be run by only one thread in a team.

Synchronization of execution of all threads can be performed by a barrier in a parallel block. It can be done by inserting

```
#pragma omp barrier
```

into the code.

Furthermore, it needs to be emphasized that OpenMP supports relaxed views of particular threads on shared variables. Specifically, after a variable has been updated by a thread, it needs to be flushed to memory so that another thread could flush its view and read the value of the variable. In other words, the flush operation makes the local view of a thread consistent with actual value in memory. The syntax of the operation is as follows:

```
#pragma omp flush [list_of_variables]
```

and performs a flush either using the specified list of variables or the whole view of a thread if a list has not been specified.

The flush operation is also called by default in many points of code that in fact synchronize threads. Some of these include (a full list can be found in [127]):

- beginning and end of `parallel` and `critical` regions,
- beginning and end of `atomic`,
- for `omp_set_lock()` and `omp_unset_lock()`.

4.2.5 Important environment variables

There are several environment variables that can help tune/influence program execution. It may be desirable to set these outside a program, depending on a specific environment:

- `OMP_NUM_THREADS val[,val1,val2,...,valn]` – sets the number of threads that can be used within parallel regions. A list of values (or just 1) separated by commas can be provided that will correspond to particular nested levels.
- `OMP_SCHEDULE schedule_type[,chunk_size]` – sets scheduling options for iterations of loop directives. `schedule_type` can be one of the following: `static`, `dynamic`, `guided`, `auto` with the meaning described in Section 4.2.2.
- `OMP_DYNAMIC val` – `val` can be either `true` or `false`. In the former case, the runtime system can dynamically set the number of threads for execution of parallel regions.
- `OMP_NESTED val` – `val` can be either `true` or `false`. In the former case, nested parallelism is possible.
- `OMP_MAX_ACTIVE_LEVELS val` – `val` denotes the maximum number of levels for nested parallel regions.
- `OMP_PLACES desc` – describes places available for execution of an application. `desc` can use either names or values, the latter typically corresponding to hardware threads. Names that can be used include: `threads` (a place denotes a hardware thread), `cores` (a place denotes a core), `sockets` (a place denotes a socket). An example of definition of places using integer values can be `{0,1,2,3,4,5},{6,7,8,9,10,11}`.
- `OMP_PROC_BIND desc` defines binding of threads during application execution. Specifically, `desc` can have the following values:
  - `true` in which case the runtime system attaches threads to places or if threads are not bound to places (`false` value),
  - a list of values separated by commas that define bound methods (thread affinity) for particular nested levels. Allowed values include [127]:
    - `close` – threads in a team are bound to places close to where the parent was placed,
    - `spread` – threads in a team are scattered among places in the parent's partition,
    - `master` – threads in a team are bound to the same place as the master thread.

– `OMP_THREAD_LIMIT val` – sets the maximum number of threads in an OpenMP application. If not specified then there is no implicit bound.

4.2.6 A sample OpenMP application

File `lnx_OpenMP_0.c` includes a C+OpenMP code for the algorithm described in Section 4.1.13. Listing 4.6 presents an important part of such an implementation. It can be seen that OpenMP really allows for very fast parallelization using OpenMP directives. This code does not explicitly use variables for thread identifiers nor the number of threads. In fact, parallelization is done automatically within the **`#pragma omp parallel for`** construct. By default, index variables within loops are private. Furthermore, since variable sum is specified in the reduction part, a private copy is created for each thread. Finally reduction is applied i.e. values from threads will be added to the `sum` variable of the master thread.

Listing 4.6 Basic OpenMP implementation of computation of $\ln(x)$

```
  int i;
  double x;
  long maxelemcount=10000000;
  double sum=0; // final value
  long power; // acts as a counter

  (...)
  // read the argument
  x=...

  // read the second argument
  maxelemcount=...

  sum=0;
  // use the OpenMP pragma for construct

#pragma omp parallel for shared(maxelemcount,x) reduction(+:sum)
  for (power=1;power<maxelemcount;power++)
    sum=sum+pow(((x-1)/x),power)/power;

  printf("\nResult is %f\n",sum);
```

The code can be compiled and run as follows using a `gcc` compiler:

```
gcc <flags> -fopenmp lnx_OpenMP_0.c -o lnx -lm
./lnx 10
```

This implementation, while simple and easy and fast to develop, has some drawbacks such as direct computation of each element of the sum in every iteration. This is important because it makes computations of various iterations independent.

These can be changed towards a longer implementation, included in file `lnx_OpenMP_1.c` resembling the MPI solution from Section 4.1.13 but implemented using a `#pragma omp parallel` in which each thread runs its own independent loop. Listing 4.7 shows an important part of such an implementation.

Listing 4.7 Improved version of an OpenMP implementation of computation of $\ln(x)$

```
  int i;
  double x;
  int threadnum,mythreadid; // the number of threads and my
    thread id
  long maxelemcount=10000000;
  double mult_coeff; // by how much multiply in every iteration
  double sum=0; // final value
  double prev; // temporary variable
  long power; // acts as a counter
  long count; // the number of elements

  (...)
  // read the argument
  x=...

  // read the second argument
  maxelemcount=...

  // use the OpenMP pragma parallel construct

#pragma omp parallel default(shared) private(power,prev,
    mythreadid,threadnum,mult_coeff,count) reduction(+:sum)
  {
    // now check the number of threads

    mythreadid=omp_get_thread_num();
    threadnum=omp_get_num_threads();

    mult_coeff=(x-1)/x;
    count=maxelemcount/threadnum;

    // now compute my own partial sum in a loop
    power=mythreadid*count+1;
```

```
    prev=pow(((x-1)/x),power);
    for(;count>0;power++,count--) {
      sum+=prev/power;
      prev*=mult_coeff;
    }
  }

  // now the main thread should have the result in variable sum
  printf("\nResult is %f\n",sum);
  (...)
```

Results for various numbers of threads per process (coded with OpenMP) are presented in Figure 4.3. It can be seen that within a node this code scales slightly better than an MPI based version presented in Section 4.1.13 which is expected within one node. Best times out of three runs for each configuration are presented.

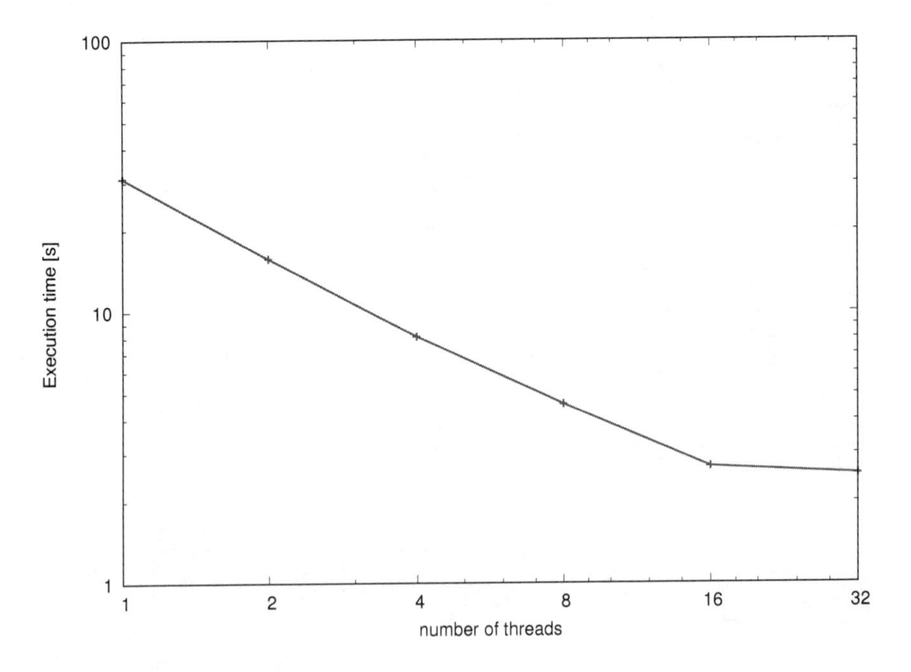

FIGURE 4.3 Execution time of the testbed OpenMP application run on a workstation with 2 x Intel Xeon E5-2620v4, 128 GB RAM, 16 physical cores, 32 logical processors, parameters: 300236771.4223 20000000000

The code was run as follows:

```
export OMP_NUM_THREADS=<number>
```

```
./a.out 300236771.4223 20000000000
```

4.2.7 Selected SIMD directives

OpenMP 4.0+ allows specification of potential execution using SIMD instructions i.e. instructions that can operate on many data elements, per specification [127]. The syntax of the basic `#pragma omp simd` directive is as follows:

```
#pragma omp simd [clauses]
<for loop>
```

which indicates that iterations of the for loop can be executed concurrently with SIMD instructions. In particular, there are a few clauses that can be used with `simd`. These include in particular: `safelen(iterdistance)` which specifies that the iteration numbers of two iterations executed concurrently must not exceed `iterdistance`, as well as `private(...)`, `lastprivate(...)` and `reduction(...)`.

Additionally, the following can be used before function:

```
#pragma omp declare simd [clauses]
<function>
```

for enabling versions of the function to process multiple arguments concurrently when the function is launched from a loop with `simd`.

Furthermore, it is possible to parallelize processing of a for loop among threads and then use SIMD instructions for processing of iterations within a thread. This can be done with (assuming many threads have already been launched):

```
#pragma omp for simd [clauses]
<for loop>
```

4.2.8 Device offload instructions

OpenMP supports launching computations on devices. This can allow starting a part of computations on e.g. coprocessors such as Intel Xeon Phi while still running parallel code on the multicore CPU(s) of the host. The key OpenMP construct allowing such offload is `#pragma omp target ...`. Specifically, the syntax of this directive is as follows:

```
#pragma omp target [clause(s)]
<block>
```

where the allowed clause includes, in particular: `private(...)`,`firstprivate (...)` lists, the `nowait` clause in which case execution of the block can be performed asynchronously. The `if (condition)` clause, in case `condition` is false will instruct the execution of the block on a host. Furthermore, `map([...]list)` is allowed which specifies mapping of variables in an original context to the device context. Variable in a list may include array (parts) and elements of structure types. If an item is not present on a device, then it is created on a device and a reference counter to it is set to 0 and then increased by 1. Mapping types can be defined before a list of variables [127]: `to` and `tofrom` copy a value of a source variable to the target one if the reference counter is 1. `alloc` would result in an undefined value in this case. Upon return from the block, `release`, `tofrom` or `from` would decrease the variable reference counter and `delete` will set it to 0. In case the reference counter reaches 0 then `tofrom` or `from` will assign the value of the variable to the original one on a host. Other cases can be found in [127].

Listing 4.8 presents an example that starts parallel processing on the device. Thanks to the `#pragma omp parallel`, threads working in parallel are launched on the device and can perform computations in parallel. Each of the threads can fetch the number of threads in the team and its unique identifier as described in Section 4.2.2.

Listing 4.8 Example using `#pragma omp target` in OpenMP

```
#include <stdio.h>
#include <omp.h>

int main(int argc,char **argv) {

  float tab0[1000];

  <initialize tab0>

#pragma omp target map(...)
  {
#pragma omp parallel ...

    ...

  }

  return 0;
}
```

As indicated in [134], offloading computations from many host threads is possible along with implicit synchronization of an OpenMP parallel region.

4.2.9 Tasking in OpenMP

OpenMP supports so-called tasking [165] that makes dynamic generation of work and assignment to threads for processing easy. Specifically, a set of tasks can be created and assigned to a group of threads running a parallel region automatically. Listing 4.9 shows an example of assignment of `tasknum` tasks to `threadnum` threads of a parallel region. File `openmp-tasking1.c` includes a complete source code of such a solution.

Listing 4.9 Example using `#pragma omp task` in OpenMP

```
  (...)
#pragma omp parallel num_threads(threadnum)
  {
    #pragma omp single
    {
      for(int i=0;i<tasknum;i++)
        {
#pragma omp task firstprivate(i)
          {
            int threadid=omp_get_thread_num();
            printf("\nTask id=%d Thread id=%d",i,threadid);
            fflush(stdout);
          }
        }
    }
  }
  (...)
```

Sample compilation using the `gcc` compiler and running the application is shown below:

```
$ gcc -fopenmp openmp-tasking1.c -o openmp-tasking1
$ ./openmp-tasking1

Task id=0 Thread id=7
Task id=2 Thread id=4
Task id=3 Thread id=5
Task id=10 Thread id=5
Task id=5 Thread id=0
Task id=12 Thread id=0
Task id=1 Thread id=1
Task id=14 Thread id=1
Task id=15 Thread id=1
Task id=13 Thread id=0
Task id=9 Thread id=4
Task id=11 Thread id=5
```

```
Task id=6 Thread id=2
Task id=8 Thread id=7
Task id=4 Thread id=3
Task id=7 Thread id=6
```

Additionally, OpenMP includes a `taskwait` construct that instructs the given task to wait for completion of all tasks started before this call [127]. Usage is as follows:

```
#pragma omp taskwait
```

4.3 PTHREADS

4.3.1 Programming model and application structure

POSIX threads or Pthreads [155] defines an API that allows writing multi-threaded programs, in particular defining constants, types and functions for:

- thread creation and waiting for other threads (joining),
- mutexes that allow implementation of mutually exclusive execution among threads,
- condition variables that are useful when a thread might need to wait for a certain condition to be satisfied by another thread.

An application needs to include file `pthread.h` in order to be able to call Pthreads functions, preceded with `pthread_`. A new thread can be created with a call to function:

```
int pthread_create(pthread_t *thread,
const pthread_attr_t *attributes,
void *(*function)(void *),void *argument)
```

with the following parameters:

- `thread` – stores an identifier of a newly created thread,
- `attributes` – attributes with which a new thread is created, can be `NULL`,
- `function` – the newly created thread starts execution of function `function`,
- `argument` – argument passed to function `function`.

Function:

```
void pthread_exit(void *value)
```

can be used to finalize the calling thread and pass a result to a thread that has called `pthread_join()` if the calling thread is joinable.

A thread can be joinable or detached. Setting a thread as a joinable thread can be done with setting a proper attribute when creating a thread i.e.:

```
pthread_t thread;
int startValue;
pthread_attr_t attr;

pthread_attr_init(&attr);
pthread_attr_setdetachstate(&attr, PTHREAD_CREATE_JOINABLE);

pthread_create(&thread,&attr,function,(void *)(&startValue));
```

The original thread can wait for another thread (join with it) by calling function:

```
int pthread_join(pthread_t thread, void **value)
```

with the following parameters:

- `thread` – identifier of a thread to join with,
- `value` – if not NULL, then output from the other thread from `pthread_exit()` will be copied to `*value`.

In case of a success, 0 is returned by the function, otherwise an error indicating either an incorrect thread identified (`ESRCH`) or a non-joinable thread (`EINVAL`). When function `function` finishes execution and returns with a result, the effect is analogous to termination with `pthread_exit()` with `result` as an argument. The attribute object is destroyed with function `pthread_attr_destroy()`. Finally, a basic structure of Pthreads code is shown in Listing 4.10.

Listing 4.10 Sample basic Pthreads structure with numbers of threads and function calls

```
#include <pthread.h>

int threadnum=2; // default
pthread_t *thread;
pthread_attr_t attr;

int *startValue;
double *threadResults; // required type can be used
```

```
void *Calculate(void *args) { // code executed by threads
    started with pthread_create()
  int start=*((int *)args); // start from this number

  // perform computations

  threadResults[start]=partialsum;
}

int main(int argc, char **argv) {
  int i;

  // read input arguments
  // the number of threads threadnum
  // can also be read from command line
  if (argc>1)
    threadnum=atoi(argv[1]);

  // allocate memory for arrays thread, startValue and
    threadResults

  // now start the threads in each process
  // define the thread as joinable
  pthread_attr_init(&attr);
  pthread_attr_setdetachstate(&attr, PTHREAD_CREATE_JOINABLE);

  for (i=0;i<threadnum;i++) {
    // initialize the start value
    startValue[i]=i;
    // launch a thread for calculations
    pthread_create(&thread[i],&attr,Calculate,(void *)(&(
    startValue[i])));
  }

  // now synchronize the threads
  for (i=0;i<threadnum;i++) {
    pthread_join(thread[i], &threadstatus);
    // merge current result with threadResults[i]
  }
  pthread_attr_destroy(&attr);
}
```

4.3.2 Mutual exclusion

The Pthreads API enables the use of mutual exclusion variables (mutexes) that allow implementation of critical sections, executed by one thread at a time. Within such a code section, a thread would usually read and modify certain variables. A mutex controls access to such a region i.e. it is locked before it and unlocked after execution of such a code section. By definition, only one thread at a time will be able to lock a mutex, the other threads would wait until it is unlocked. The Pthreads API allows initialization of a mutex using the following function:

```
int pthread_mutex_init(pthread_mutex_t *mutex,
const pthread_mutexattr_t *mutexattributes)
```

with the following parameters:

- `mutex` – address of a mutex object,
- `mutexattributes` – attributes, can be NULL for default.

In case of successful initialization, the value of 0 is returned. Otherwise, certain codes indicate an error, in particular `EINVAL` – incorrect attributes, `EBUSY` – mutex already initialized, `ENOMEM` – not enough memory.

There can be a few types of mutexes that will determine behavior of subsequent mutex locking or unlocking [155]. Subsequent locking is considered here as locking of an already locked mutex, subsequent unlocking as unlocking an already unlocked mutex or unlocking a mutex locked by another thread.

Additionally, a robust mutex is a mutex that can be useful when a thread that has locked a mutex finishes. In such a case, another thread waiting to lock the mutex receives the value of `EOWNERDEAD`. Making a mutex robust can be done with a call to function:

```
int pthread_mutexattr_setrobust(pthread_mutexattr_t *attributes,
    int robustflag)
```

where, apart from the `attributes parameter`, `robustflag` indicates whether a mutex (to which attributes will be assigned during initialization) will be robust. In order to achieve this, `robustflag` should be set to `PTHREAD_MUTEX_ROBUST` (instead of the default `PTHREAD_MUTEX_STALLED`). The value of 0 indicates a success while `EINVAL` an incorrect flag. Furthermore, after such termination the mutex is inconsistent. Its state can be marked as consistent with a call to function:

```
int pthread_mutex_consistent(pthread_mutex_t *mutex)
```

The value of 0 indicates a success while `EINVAL` an error (cannot make the given mutex as consistent in this case).

Mutex types include:

- PTHREAD_MUTEX_DEFAULT – it can either correspond to one of the following types or be undefined for subsequent locking or subsequent unlocking for a mutex that is not robust. For a robust mutex, it can either correspond to one of the following types or be undefined for subsequent locking or return an error for subsequent unlocking.
- PTHREAD_MUTEX_NORMAL – subsequent locking a normal mutex results in a deadlock while subsequent unlocking a robust mutex results in an error and subsequent unlocking a mutex that is not locked results in undefined behavior.
- PTHREAD_MUTEX_RECURSIVE – there is an integer count associated with a recursive mutex. Initially set to 0, it is increased by 1 when a mutex is locked or decreased by 1 when unlocked. The value of 0 allows threads to lock such a mutex. Subsequent unlocking of a recursive mutex results in an error.
- PTHREAD_MUTEX_ERRORCHECK – in this case, either subsequent locking or subsequent unlocking result in errors.

Functions:

```
int pthread_mutexattr_init(pthread_mutexattr_t *attribute)
int pthread_mutexattr_settype(pthread_mutexattr_t *attribute,
    int type)
int pthread_mutexattr_destroy(pthread_mutexattr_t *attribute)
```

can be used for attribute manipulation.

After a mutex has been used, it can be released with a call to function:

```
int pthread_mutex_destroy(pthread_mutex_t *mutex)
```

Similarly to the previous function, the value of 0 is returned in case of a success, otherwise EINVAL – indicates an incorrect mutex while EBUSY – indicates mutex is already initialized and still locked or used. After a mutex has been initialized, it can be used. Specifically, locking a mutex is performed with a call to function:

```
int pthread_mutex_lock(pthread_mutex_t *mutex)
```

After a call has returned, the calling thread locked the mutex. The calling thread will block if the given mutex is already locked.

Unlocking a given mutex can be performed with a call to the following function:

```
int pthread_mutex_unlock(pthread_mutex_t *mutex)
```

Function:

```
int pthread_mutex_trylock(pthread_mutex_t *mutex)
```

as its name suggests, will lock the mutex if it is unlocked or will return at once otherwise. For a mutex of type `PTHREAD_MUTEX_RECURSIVE` already locked by the calling thread, it will increase the corresponding count.
Function:

```
int pthread_mutex_timedlock(pthread_mutex_t *restrict mutex,
const struct timespec *restrict timeout)
```

locks the `mutex` if it is possible or otherwise waits for the mutex to be unlocked. The waiting time is determined by `timeout`.

Locking and unlocking functions return 0 if successful. In case of `pthread_mutex_trylock()` the value of 0 is returned in case the mutex has been locked by the call.

4.3.3 Using condition variables

Condition variables, used along with mutexes, allow more complex synchronization among threads in which waiting on conditions depending on other threads is required. An example could be implementation of the producer-consumer problem.
Function:

```
int pthread_cond_init(pthread_cond_t *restrict conditionvar,
const pthread_condattr_t *restrict attributes)
```

with the following parameters:

- `conditionvar` – a pointer to a condition variable to initialize,
- `attributes` – attributes for initialization, can be NULL for default values,

initializes the given condition variable. Destruction or deinitialization of condition variable `conditionvar` is performed with a call to function:

```
int pthread_cond_destroy(pthread_cond_t *conditionvar)
```

Both functions return 0 in case of success. `ENOMEM` during initialization indicates lack of memory and `EAGAIN` lack of other resources.
Function:

```
int pthread_cond_wait(pthread_cond_t *restrict conditionvar,
pthread_mutex_t *restrict mutex)
```

with the following parameters:

- `conditionvar` – a condition variable,
- `mutex` – a mutex,

forces a thread to wait on a given condition variable. It should be called from within a locked mutex which is released then.
Function:

```
int pthread_cond_timedwait(pthread_cond_t *restrict
conditionvar,pthread_mutex_t *restrict mutex,
const struct timespec *restrict absolutetime)
```

acts as a wait until the given `absolutetime`. If the system time passes `absolutetime`, the function returns `ETIMEDOUT`. `EINVAL` denotes an invalid absolute time.
Function:

```
int pthread_cond_signal(pthread_cond_t *conditionvar)
```

will cause unblocking of one or more threads blocked on the pointed condition variable while function:

```
int pthread_cond_broadcast(pthread_cond_t *conditionvar)
```

will cause unblocking of all threads blocked on the pointed condition variable. Successful termination returns the value of 0.

4.3.4 Barrier

Function:

```
int pthread_barrier_init(pthread_barrier_t *restrict barrier,
const pthread_barrierattr_t *restrict attributes,
unsigned threadcount)
```

initializes required resources needed to use a barrier pointed at by `barrier`. The other parameters include:

- `attributes` – attributes (NULL indicates default values),
- `threadcount` – indicates the number of threads that form the barrier i.e. which are expected to call function `pthread_barrier_wait()` so that the barrier is completed and the threads are free to continue.

Function:

```
int pthread_barrier_destroy(pthread_barrier_t *barrier)
```

releases resources used by the barrier pointed at by `barrier`. Both functions return 0 in case of success. `ENOMEM` indicates lack of memory, `EAGAIN` lack of other resources while `EINVAL` an invalid value of `threadcount`.
Function:

```
int pthread_barrier_wait(pthread_barrier_t *barrier)
```

can be used by threads to enter the barrier. After `threadcount` threads have entered the barrier, all are allowed to continue. Then the function returns `PTHREAD_BARRIER_SERIAL_THREAD` for one of the threads and the value of 0 for the other threads.

4.3.5 Synchronization

An application should not allow a thread to access a variable while there may be modifications to the variable performed by another thread. According to the specification, there are several functions that synchronize memory view among threads. Among the functions described in this book, there are: `pthread_create()`, `pthread_join()`, `pthread_mutex_lock()` (with the exception of a recursive mutex already locked by the thread that has called the function), `pthread_mutex_trylock()` `pthread_mutex_timedlock()`, `pthread_mutex_unlock()` (with the exception of a recursive mutex with a counter larger than 1), `pthread_cond_wait()`, `pthread_cond_timedwait()`, `pthread_cond_signal()` and `pthread_cond_broadcast()`, `pthread_barrier_wait()`. Specification [155] lists additional functions, not covered in this book.

4.3.6 A sample Pthreads application

Following previously presented implementations of parallel computation of a sum, this section presents a Pthreads implementation.
Function `Calculate()`:

```
void *Calculate(void *args) {
  int start=*((int *)args); // start from this number
  double partialsum=0; // partial sum computed by each process
  double mult_coeff; // by how much multiply in every iteration
  double prev; // temporary variable
  long power; // acts as a counter
  long count; // the number of elements

  // each process performs computations on its part
```

```
  mult_coeff=(x-1)/x;
  count=maxelemcount/threadnum;
  // now compute my own partial sum in a loop
  power=start*count+1;
  prev=pow(((x-1)/x),power);
  for(;count>0;power++,count--) {
    partialsum+=prev/power;
    prev*=mult_coeff;
  }
  threadResults[start]=partialsum;
}
```

is executed by each of several threads launched by the main thread using `pthread_create()`. The most important code fragment of the latter thread is shown in Listing 4.11.

Listing 4.11 Parallel implementation of computing $\ln(x)$ using Pthreads

```
(...)

int main(int argc, char **argv) {

  int i;
  void *threadstatus;

  // check if needed arguments were provided
  (...)

  // read the argument
  x=...

  // read the second argument
  maxelemcount=...

  // read the third argument if provided (otherwise the
  // default number of threads
  threadnum=...

  (...)

  thread=(pthread_t *)malloc(sizeof(pthread_t)*threadnum);
  if (thread==NULL) {
    // handle error
  }

  startValue=(int *)malloc(sizeof(int)*threadnum);
```

```
    if (startValue==NULL) {
      // handle error
    }

    threadResults=(double *)malloc(sizeof(double)*threadnum);
    if (threadResults==NULL) {
      // handle error
    }

    // now start the threads
    // define the thread as joinable
    pthread_attr_init(&attr);
    pthread_attr_setdetachstate(&attr, PTHREAD_CREATE_JOINABLE);

    for (i=0;i<threadnum;i++) {
      // initialize the start value
      startValue[i]=i;
      // launch a thread for calculations
      pthread_create(&thread[i], &attr, Calculate, (void *)(&(
      startValue[i])));
    }

    // now synchronize the threads
    // and add results from all the threads
    for (i=0;i<threadnum;i++) {
      pthread_join(thread[i], &threadstatus);
      totalsum+=threadResults[i];
    }
    printf("Result=%f\n",totalsum);
    pthread_attr_destroy(&attr);
    return 0;
}
```

The program can be compiled and run as follows, as an example:

```
gcc <flags> lnx_Pthreads.c -lpthread -lm
./a.out 1.2 1000000000 16
```

Figure 4.4 presents execution times of the application run on a workstation with 2 x Intel Xeon E5-2620v4, 128 GB RAM, for various numbers of threads.

4.4 CUDA

4.4.1 Programming model and application structure

NVIDIA CUDA is an API that allows parallel programming for a GPU or a collection of GPUs. It proposes a programming model that exposes massive

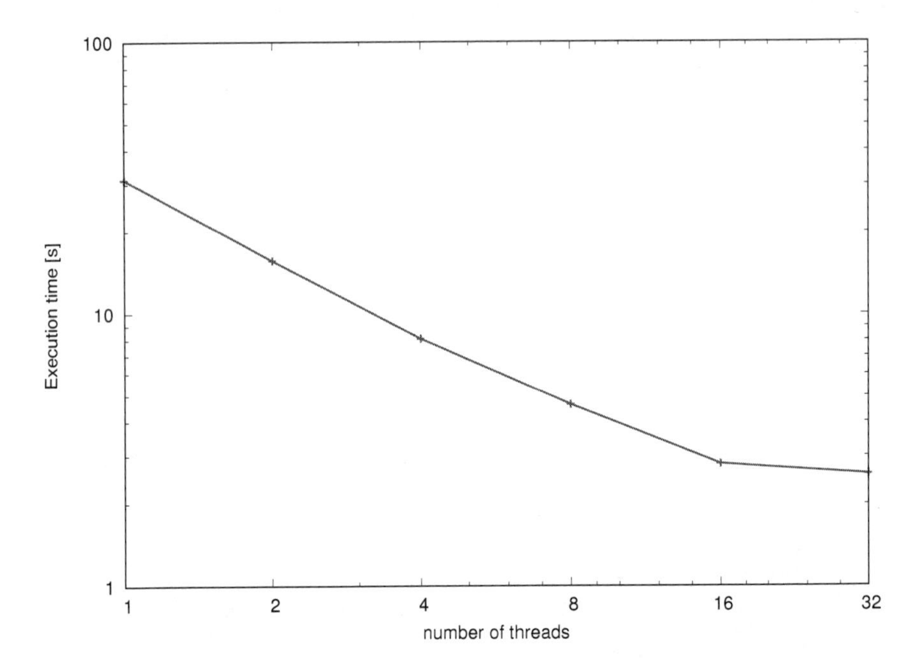

FIGURE 4.4 Execution time of the testbed Pthreads application run on a workstation with 2 x Intel Xeon E5-2620v4, 128 GB RAM, 16 physical cores, 32 logical processors, parameters: 300236771.4223 20000000000

parallelism of a potentially very large number of threads running on streaming multiprocessors (SMs) of a GPU. The most efficient type of code to execute on a GPU is the one in which threads execute the same instructions on various input data.

The structure of parallel code running on a GPU is shown in Figure 4.5. It distinguishes the following components:

- grid – represents the whole application and consists of a number of thread blocks,
- thread blocks – each of these groups a certain number of threads.

In fact, a standard application starts a process on a CPU that follows these steps:

1. Allocate memory using `malloc(...)` and initialize data in RAM.
2. Allocate memory on a GPU using `cudaMalloc(...)` for input/output data.

3. Copy input data from RAM to the memory on the GPU using `cudaMemcpy(...)`.
4. Launch computations on the GPU by calling a function called a *kernel function*. Such a function is declared using the `__global__` qualifier and has `void` as the returned type.
5. Upon completion copy results from the memory on the GPU to RAM using `cudaMemcpy(...)`.
6. Display results.

An application may call several kernel functions in a sequence. Additionally, it is also possible to manage computations on several GPU cards by running many kernels as shown in Section 4.4.8.

Running a kernel starts computations on a GPU with threads arranged in blocks and blocks in a grid, as shown in Figure 4.5. This requires setting the configuration of a grid when calling a kernel function. The syntax is as follows:

```
nameofkernelfunction<<numberofblocksingrid,
numberofthreadsinblock>>(kernelparameters);
```

Specifically, depending on the programmer, threads in a block and blocks in a grid can be arranged in 1, 2 or 3 dimensions. Typically, this will correspond to the number(s) of dimensions used in the algorithm itself for natural assignment of threads to problem data. For instance:

- operations on vectors would naturally map to 1 dimension in which each thread can be assigned a given element of vectors to perform an operation on,
- multiplication of two matrices would map to 2 dimensions,
- simulation of winds in a valley would map to 3 dimensions.

When using one dimension, variables `numberofblocksingrid`, `numberofthreadsinblock` can be of type `int`. In general type `dim3` can be used for specification of sizes in more dimensions as follows:

```
dim3 numberofblocksingrid(32,32,32);
dim3 numberofthreadsinblock(16,16,4);
```

Each thread would typically find its location in the grid in order to find out the data/part of computations it should perform. The following variables can be used for this purpose:

- `threadIdx.<dim>` – id of a thread within a block, where `<dim>` can be `x`, `y` or `z` depending on the dimension referred to,

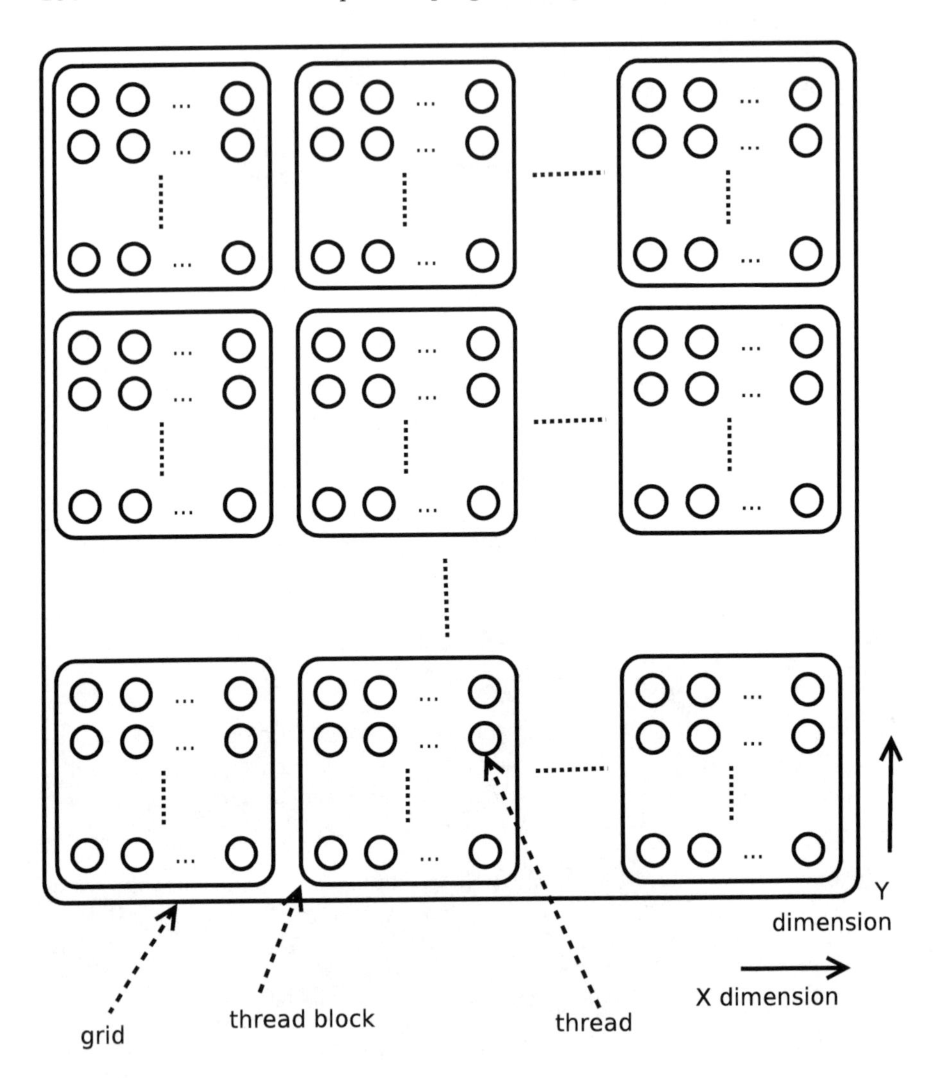

FIGURE 4.5 Architecture of a CUDA application, a 2D example

- `blockIdx.<dim>` – id of a block within a grid, where `<dim>` can be `x`, `y` or `z` depending on the dimension referred to,
- `blockDim.<dim>` – size of a block in threads, where `<dim>` can be `x`, `y` or `z` depending on the dimension referred to,
- `gridDim.<dim>` – size of a grid in blocks, where `<dim>` can be `x`, `y` or `z` depending on the dimension referred to.

In CUDA, several function scope related qualifiers are available, including [122]:

- `__device__` – denotes a function executed and accessible only from a device,
- `__global__` – denotes a function executed on a device and accessible from a host and latest devices supporting dynamic parallelism,
- `__host__` – denotes a function executed on a host and accessible only from the host.

Types of memories available on a GPU are shown in Figure 4.6.
In CUDA, several data related qualifiers are available, including [122]:

- `__device__` – variable to be stored in global memory on a device,
- `__constant__` – variable to be stored in constant memory on a device,
- `__shared__` – variable to be stored in shared memory on a device – visible to threads of a relevant block,
- `__managed__` – variable that can be accessed from the host and the device.

4.4.2 Scheduling and synchronization

Threads are lightweight and, compared to threads typically launched on a CPU, may be assigned relatively few computations which can still result in high performance. A programmer does not explicitly assign threads to system processors and cores. Instead, the runtime system does that. By design, execution of blocks is independent. On the other hand, execution of threads within a single block can be synchronized. This is achieved by calling function:

```
__syncthreads()
```

Typically, `__syncthreads()` will be used when all threads in a block perform the following sequence of operations:

1. Each thread copies part of input data from global to shared memory (shared memory can be used as kind of cache as discussed in Section 6.9.1).
2. `__syncthreads()` is called that ensures that all the threads have completed copying.
3. Each thread computes part of data and stores in shared memory.
4. `__syncthreads()` is called that ensures that all the threads have written their parts of data.

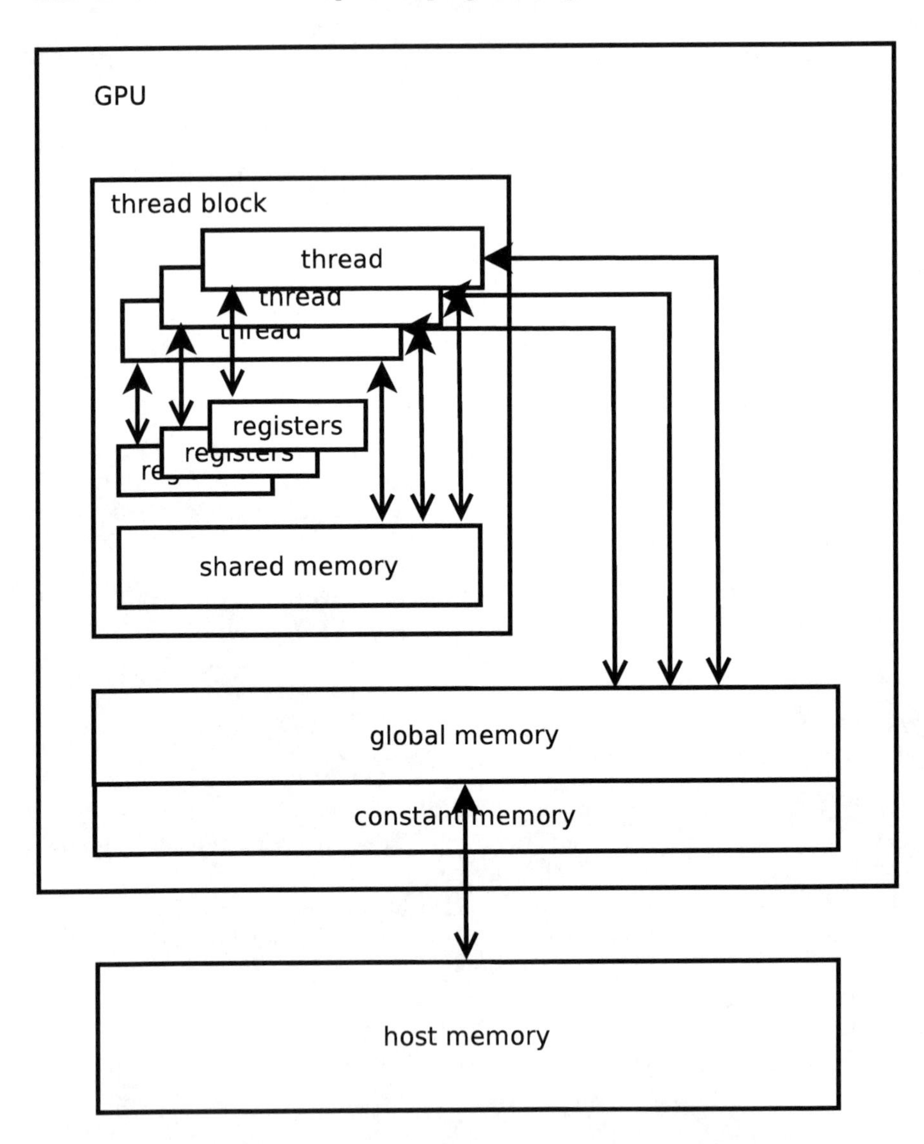

FIGURE 4.6 Memory types and sample usage by an application running on a GPU

5. Data is used by threads for subsequent computations (some threads may use output computed by other threads).

It should be noted that function `__syncthreads()` must be called by all threads within a block. Scheduling is performed by the runtime layer which uses groups of 32 threads called *warps* as smallest scheduling components.

Having independent blocks allows the runtime layer to hide latency in case some delays show up. From the programmer's point of view, this results in automatic dynamic scheduling on the GPU and load balancing but also imposes constraints on arrangement of computations and data in the application.

The programming model for an application is called SIMT (Single Instruction Multiple Threads). Essentially, threads working in parallel usually process various data or perform various computations, the results of which are further integrated into a final result. Threads within a warp must execute the same instruction at the same time. It should be noted, however, that if the code includes conditional statements such as:

```
if (condition) {
  <code of block A>
else {
  <code of block B>
}
```

then codes of blocks A and B will not be executed in parallel if some threads have `condition` equal to `true` and others equal to `false`. In fact, codes of blocks A and B will be executed sequentially. Starting with Volta and CUDA 9, independent thread scheduling is used with program counter per thread and statements from divergent branches can be interleaved [65]. Function `__syncwarp()` can be used for reconvergence. Additionally, Cooperative Groups in CUDA 9 allow to manage groups of communicating threads in a flexible way with respective functions [122] and possibility of whole grid or multi-GPU synchronization.

CUDA offers several so-called atomic functions that allow atomic operations that read, modify and store values in either global or shared memories [122]. Such operations can be called from within functions running on a device. Particular functions take the input value and perform an operation using the old value at the given address and the new value and saves a result to the same location the old value was stored. Examples of useful functions include: `int atomicAdd(int *oldvaluelocation,int value)` and `int atomicSub(int *oldvaluelocation,int value)` for adding or subtracting `value` from a variable stored in a given location, `int atomicMax(int *oldvaluelocation,int value)` and `int atomicMin(int *oldvaluelocation,int value)` for computing a maximum or minimum out of the old value and newly given `value`, `int atomicExch(int *oldvaluelocation,int value)` for exchanging the old and new values as well as `int atomicCAS(int *oldvaluelocation,int valuetocompare,int valueifequal)` that compares the old value with `valuetocompare` and stores `valueifequal` if compare returned `true` or does not do anything otherwise. CUDA also supports atomic logical operations.

4.4.3 Constraints

When processing on a GPU, there are several constraints that should be taken into account, in particular:

1. number of threads per block is limited, currently to 1024 for compute capability 2+,
2. number of resident threads per streaming multiprocessor,
3. number of registers per thread, number of registers per block, number of registers per multiprocessor,
4. number of resident blocks per multiprocessor,
5. amount of shared memory per multiprocessor.

It should be noted that requirements on shared memory and registers of threads within a block may be such that the runtime system may not be able to run as many blocks in parallel as potentially possible from one of the above requirements. This can potentially limit parallelism.

Compute capability of a GPU device refers to functions available on the device and specifications, including constraints.

4.4.4 A sample CUDA application

File `collatz.cu` includes a parallel CUDA code for finding the maximum number of iterations needed to reach 1 in the Collatz hypothesis sequence across a range of input integer values. The Collatz hypothesis states that starting with any positive integer number it will eventually reach 1 when performing the following operations:

1. if the number is even, divide it by 2,
2. otherwise multiply the number by 3 and add 1.

Then the same approach is adopted to a result and the procedure is repeated. Listing 4.12 presents an important part of such an implementation. The code creates a grid with blocks in which each thread deals with a separate input argument. Afterwards, each thread writes the number of iterations for its number into its own space in a global array. Note that the code also runs for `my_index=0` but this does not affect the final result.

Listing 4.12 Sample CUDA implementation of a program for checking the Collatz hypothesis

```
(...)
```

```
__global__
void checkCollatz(long *result) {
  long my_index=blockIdx.x*blockDim.x+threadIdx.x;
  unsigned long start=my_index;
  char cond=1;
  unsigned long counter=0;

  for(;cond;counter++) {
    start=(start%2)?(3*start+1):(start/2);
    cond=(start>1)?1:0;
  }

  result[my_index]=counter;
}

int main(int argc,char **argv) {

  struct timeval start,stop,start1,stop1;
  long result;
  int threadsinblock=1024;
  int blocksingrid=10000;
  long threadcount=threadsinblock*blocksingrid;

  (...)

  long size=threadcount*sizeof(long);
  long *hresults=(long *)malloc(size);
  if (!hresults) errorexit("Error allocating memory on the host"
    );

  long *dresults=NULL;
  if (cudaSuccess!=cudaMalloc((void **)&dresults,size))
    errorexit("Error allocating memory on the GPU");

  (...)

  // start computations on the GPU
  checkCollatz<<<blocksingrid,threadsinblock>>>(dresults);
  if (cudaSuccess!=cudaGetLastError())
    errorexit("Error during kernel launch");

  if (cudaSuccess!=cudaMemcpy(hresults,dresults,size,
    cudaMemcpyDeviceToHost))
     errorexit("Error copying results");
```

```
  cudaDeviceSynchronize();

  // now find the maximum number on the host
  result=0;
  for(long i=0;i<threadcount;i++)
    if (hresults[i]>result)
      result=hresults[i];

  printf("\nThe final result is %ld\n",result);
  (...)

  // release resources
  free(hresults);
  if (cudaSuccess!=cudaFree(dresults))
    errorexit("Error when deallocating space on the GPU");
}
```

This initial code can be improved in several ways. For instance, this code does not exploit shared memory that could be used within thread blocks. As a side effect, the size of the result that needs to be sent back to the host memory is large and consequently finding a maximum on the host side can take considerable time. An improved version, included in file `collatz-partialparallelmaxcomputation.cu` applies parallel reduction within each thread block using shared memory. An improved kernel is shown in Listing 4.13. Specifically, each thread puts its result into its own place in shared memory and then a reduction algorithm proceeds in the logarithmic number of steps for finding a maximum value synchronized using `__syncthreads()` between iterations. Also, the number of results transferred to the host side and browsed sequentially for a maximum decreased as a result. This code could potentially be improved further as shown in [82].

Furthermore, another improvement could be to parallelize finding the maximum on the host side, for instance using OpenMP.

Listing 4.13 Improved CUDA implementation of a program for checking the Collatz hypothesis

```
__global__
void checkCollatz(long *result) {
  long my_index=blockIdx.x*blockDim.x+threadIdx.x;
  unsigned long start=my_index;
  char cond=1;
  unsigned long counter=0;
  __shared__ long sresults[1024];

  for(;cond;counter++) {
```

```
    start=(start%2)?(3*start+1):(start/2);
    cond=(start>1)?1:0;
  }

  sresults[threadIdx.x]=counter;
  __syncthreads();

  // now computations of a max within a thread block can take
    place
  for(counter=512;counter>0;counter/=2) {
    if (threadIdx.x<counter)
      sresults[threadIdx.x]=(sresults[threadIdx.x]>sresults[
    threadIdx.x+counter])?sresults[threadIdx.x]:sresults[
    threadIdx.x+counter];
    __syncthreads();
  }

  // store the block maximum in a global memory
  if (threadIdx.x==0)
    result[blockIdx.x]=sresults[0];
}
```

Kernel running times can be checked with:

```
nvcc collatz.cu
nvprof ./a.out
nvcc collatz-partialparallelmaxcomputation.cu
nvprof ./a.out
```

Table 4.3 presents lowest execution times for the GPU part and the whole program for a testbed platform.

TABLE 4.3 Execution times [us] for two CUDA code versions

	2 x Intel Xeon E5-2620v4 + NVIDIA GTX 1070
standard execution time GPU	54221
improved execution time GPU	23329

4.4.5 Streams and asynchronous operations

In a typical scenario a host thread allocates space on a device, copies input data to the device, invokes a kernel and copies results back to the host memory:

```
cudaMalloc(...);
```

```
cudaMemcpy(...,inputdata,...,cudaMemcpyHostToDevice);
kernel<<<gridsize,blocksize>>>(...);
cudaMemcpy(results,...,...,cudaMemcpyDeviceToHost);
```

Kernel invocation is asynchronous. According to [122] copying data between a host and a device with 64KB or smaller size or copying within one device are also asynchronous from the point of view of the calling host thread.

Consequently, it is easily possible to queue kernel invocations and perform computations on the host in parallel with processing on the device. This can be accomplished with the following code:

```
kernelA<<<gridsizeA,blocksizeA>>>(...);
kernelB<<<gridsizeB,blocksizeB>>>(...);
kernelC<<<gridsizeC,blocksizeC>>>(...);
processdataonCPU();
cudaDeviceSynchronize();
```

Function `cudaDeviceSynchronize()` will not terminate until all previously queued tasks on the devices have finished.

CUDA API offers possibilities to overlap communication between a host and a device as well as computations on the device. The following can be used for this purpose:

- Streams – operations issued to a stream are executed one after one; the default stream (with id 0) is used if no other stream is specified explicitly. Operations in various streams (with non 0 ids) can be executed concurrently.
- Asynchronous communication between a host and a device. Whether a particular device supports such mode can be checked by reading the value of field `asyncEngineCount` in a structure returned by function `cudaGetDeviceProperties(...)` described in Section 4.4.8.

A stream can be created using function:

```
cudaError_t cudaStreamCreate(cudaStream_t *stream)
```

and destroyed if not needed anymore using function:

```
cudaError_t cudaStreamDestroy(cudaStream_t stream)
```

Indication of the stream in which computations should be performed in the device is done in the kernel invocation as follows:

```
kernelA<<<gridsizeA,blocksizeA,smbytes,stream>>>()
```

where smbytes indicates dynamically reserved shared memory size per thread block [84, 122] and stream indicates the stream to be used. For the former, it is possible to specify the data size for shared memory when launching the kernel as demonstrated in the example in Section 5.1.5.

An asynchronous call starting communication between the host and a device can be performed as follows:

```
cudaError_t cudaMemcpyAsync(void *destinationbuffer,
const void *sourcebuffer,size_t size,
enum transfertype,cudaStream_t stream = 0)
```

where stream indicates the stream the request should be put into. Consequently, this allows potential overlapping of computations and communication when using two streams as shown in Section 6.1.2. Asynchronous copying and overlapping of communication between the host and the device with computations can be done when page-locked memory on the host is used. Page-locked (i.e. pinned) memory can be allocated using function:

```
cudaError_t cudaHostAlloc(void **memory,
size_t sizeinbytes,unsigned int flag)
```

and can be released using a call to function:

```
cudaError_t cudaFreeHost(void *memory)
```

If cudaHostAllocMapped is used as flag in cudaHostAlloc(...) then such memory can be referred to from a device. The latter can fetch a respective pointer with a call to function:

```
cudaError_t cudaHostGetDevicePointer(void
**pointerondevice,void *pointeronhost,
unsigned int flags)
```

where pointerondevice will return a pointer that can be used on the device [123].

CUDA allows concurrent kernel execution as outlined in [137]. There are a few conditions for that to take place:

- kernel execution must have been submitted to different streams,
- if there are GPU resources left having considered previously submitted kernels.

In CUDA 7 a new possibility for so-called per thread default streams was introduced [85]. Specifically, it can be useful for multithreaded applications in which case each thread running on the host will have its own default stream and such per thread default streams will execute tasks in parallel. Turning on per thread default streams can be done by compiling using `nvcc` with the `--default-stream per-thread` flag. Furthermore, such per thread streams will execute tasks in parallel with streams other than the default one. There are several functions that can be used for synchronization in this context:

```
cudaDeviceSynchronize(void)
```

which waits until all tasks previously submitted to the device have completed,

```
cudaError_t cudaStreamSynchronize(cudaStream_t stream)
```

which waits until tasks previously submitted to the stream have completed,

```
cudaError_t cudaStreamWaitEvent(cudaStream_t stream,
cudaEvent_t event,
unsigned int flags)
```

which forces all tasks that will be submitted to the given `stream` to wait for the given `event`. It is important that this function can be used to synchronize tasks submitted to various devices if various streams are used. Specifically, all the tasks to be submitted to the stream will not start until a call to function `cudaEventRecord(...)` has completed [123]. In order to implement such synchronization, an event needs to be created first and recorded later with function `cudaEventRecord(...)` [113, 137]. Functions that allow management of events are as follows:

```
cudaError_t cudaEventCreate(cudaEvent_t *event)
```

which creates an event with standard settings (see below for creation of an event with flags),

```
cudaError_t cudaEventDestroy(cudaEvent_t event)
```

which destroys the object representing the event,

```
cudaError_t cudaEventRecord(cudaEvent_t event,
cudaStream_t stream)
```

for asynchronous recording of an event in a stream. Consequently, if it is needed to check the status, it can be done so with function `cudaError_t cudaEventQuery(cudaEvent_t event)` which returns, in particular, `cudaSuccess` in case tasks executed before `cudaEventRecord(...)` have completed or `cudaErrorNotReady` otherwise.
Function:

```
cudaError_t cudaEventSynchronize(cudaEvent_t event)
```

enforces waiting until all tasks submitted before the last call to `cudaEventRecord(...)` have completed.
The way the host thread is waiting for results can be affected by setting proper device flags such as [123]:

- `cudaDeviceScheduleSpin` – do not yield control to other threads until receiving result from a GPU and busy wait for result that may affect performance of other threads running on the CPU,
- `cudaDeviceScheduleYield` – yield control to other threads before receiving result from a GPU – may increase the time for getting results but does not affect performance of other threads running on the CPU that much,
- `cudaDeviceScheduleBlockingSync` – block the calling thread.

Following this, function:

```
cudaError_t cudaEventCreateWithFlags(cudaEvent_t
*event,unsigned int flags)
```

allows creation of an event with additional flags. Specifically, if flag `cudaEventBlockingSync` is used then the thread waiting for completion of tasks before this event (using `cudaEventSynchronize`) will block.

Additionally, there are functions that allow for querying statuses of a stream and an event:

```
cudaError_t cudaStreamQuery(cudaStream_t stream)
```

which returns `cudaSuccess` if all tasks previously submitted to the given `stream` have already been finalized, `cudaErrorNotReady` is returned otherwise,

```
cudaError_t cudaEventQuery(cudaEvent_t event)
```

which returns `cudaSuccess` if all tasks submitted prior to the last call to `cudaEventRecord(...)` have already been finalized, `cudaErrorNotReady` is returned otherwise.

4.4.6 Dynamic parallelism

Starting with CUDA 5 and on GPUs with compute capability 3.5+, so-called dynamic parallelism has become available [98, 10, 9]. This has enabled natural and more efficient than before implementation of divide-and-conquer algorithms, including irregular ones.

Dynamic parallelism allows launching kernels from within kernels i.e. launching kernels from a code running on a GPU. This can shorten execution times of certain application compared to an implementation without dynamic parallelism [96]. What is important is that in many divide-and-conquer algorithms, whether a given problem is divided into subproblems or not is only determined at runtime. Similarly, the number of subproblems and their sizes can also be known only at runtime. Examples of such problems include:

- quicksort [35]: further partitioning of a problem is performed if the number of elements is greater than 2, the number of subproblems is 2,
- alpha beta search [38] in games such as chess: further partitioning of a problem is performed if the depth has reached a predefined threshold or end position is achieved, the number of subproblems is equal to the number of considered moves,
- adaptive quadrature integration [170, 38]: further partitioning of a problem depends on whether sufficient accuracy has been obtained by prior partitioning of a data range,

Such problems can be effectively implemented on a GPU by invoking kernels from within a kernel. From the implementation point of view, several assumptions and constraints should be noted and considered when using dynamic parallelism [10, 98]:

1. Already spawned threads of a parent kernel can launch child kernels. A thread can start execution of a child kernel. It is important to note that execution of the parent kernel function will be finalized only after execution of child kernel functions has completed.
2. Calling `cudaDeviceSynchronize()` would wait for kernel functions started, for instance if the thread needs output from a child kernel. However, the calling thread would not know about timing of kernels launched by other threads within a thread block. Consequently, synchronization among launching threads might be necessary.
3. Streams and events that were created within a thread block are shared and can be used by its threads.
4. Launching child kernels in parallel can be achieved using separate non-default streams.
5. The following fundamental rules apply to visibility of particular memory types when used with dynamic parallelism [122]:
    - Global and constant memories are shared between parent and child kernels.
    - Shared memories for parent and child kernels are different.

- A child kernel sees memory of the calling thread.
- The parent kernel sees updates of a child kernel after synchronization (waiting for the kernel to complete). Other threads of the parent kernel would see results after synchronization is achieved by calling `__syncthreads()`. Block wide synchronization in case a thread launched a child kernel would include invoking `__syncthreads()`, `cudaDeviceSynchronize()` by the thread that launched a child kernel and invoking `__syncthreads()` again.
- The maximum depth on which synchronization is performed can be set as `cudaLimitDevRuntimeSyncDepth` and should be set if it is supposed to exceed 2 [10], otherwise it will not be put into effect. It can be set using a call to `cudaDeviceSetLimit(...)` as shown in the example in Section 5.3.2. Other limits can also be set as shown in [10].

6. Streams on a GPU must be created by calling `cudaStreamCreateWithFlags()` with the `cudaStreamNonBlocking` flag. This allows spawning kernel functions concurrently (without such streams child kernels would be executed one by one).

7. Compiling and linking requires architecture 3.5 or higher and can be achieved with:

```
nvcc -arch=sm_35 -rdc=true  dynamicparallelism.cu \
-lcudadevrt
```

   where the `rdc` flag enables generation of relocatable device code.

8. The maximum depth of recursive kernel invocations is limited and is set to 24 but can be reached sooner if memory limitations show up during starting a new kernel.

In Section 5.3.2 a code for implementation of a divide-and-conquer problem is given.

4.4.7 Unified Memory in CUDA

CUDA 6 introduced the concept of so-called unified memory that allows simplification of the programming model significantly. The traditional model described above typically allocates spaces on both the host and device sides, copies input data from the host to device memory, performs computations and copies results back from the device to the host memory. Unified memory, on the other hand, allows writing code that very much resembles code running on a CPU. Specifically, it is enough to perform the following steps:

1. Allocate space in unified memory – this can be done using function:

```
cudaError_t cudaMallocManaged(void **memory,size_t size,
unsigned int flags)
```

where `memory` will point at a newly allocated memory space on the device that will be managed by Unified Memory mechanisms. The size of the memory is specified in `size` and given in bytes. It should be noted that on card generations before Pascal, pages would be populated on a GPU. If a CPU initializes data then a page(s) will be moved to the CPU and then to the GPU if a kernel operating on it is launched. On Pascal cards, no pages are created at the managed malloc, only when referenced which can minimize page migrations [143]. On Pascal, concurrent accesses from a CPU and a GPU are possible. Flags, if specified, may include `cudaMemAttachGlobal` which indicates that the newly allocated memory can be referred to from any stream on a GPU. The memory can be accessed in both CPU and GPU threads. In case there are several GPUs in the system, memory is allocated on the one that was selected when `cudaMallocManaged(...)` was called. Threads running on other GPUs may refer to the memory as well. A sample allocation of memory can be done as follows:

```
char *memory;
int memsize=1024;
cudaMallocManaged(&memory,memsize);
```

2. Initialize data.

3. Spawn a kernel on a device (it will have access to the data in the unified memory.

4. Wait for completion of computations on the host side which can be accomplished using a call to `cudaDeviceSynchronize()`.

5. Read results from the unified memory and display to the user.

Unified memory can be accessed from both threads running on a CPU and a device. It should be noted that, in general, the host thread must not access managed memory (allocated with `cudaMallocManaged(...)`) while a kernel on a GPU is active, even if the latter is not actually using the data or is using other variables [122]. This is, however, possible on Pascal generation of cards [143].

Unified memory allows increasing productivity for CUDA-enabled applications and porting traditional HPC codes to GPU enabled environments. In particular, management of data stored in complex structures is much easier using unified memory rather than explicit manual copying between the host and device memories. Work [83] shows how C++ `new`, `delete` operators as well as class constructors can be overloaded to hide using unified memory, especially allocation in unified memory and initialization.

Paper [103] analyzes performance of Unified Memory Access (UMA) vs non-UMA implementations of selected benchmarks. Specifically, the paper considers microbenchmarks that vary in access schemes to data and whether CPU/GPU sides access all or parts of data. In general, UMA generated higher overhead than the traditional approach with the exception of the scenario with Successive Over Relaxation for small data sizes. The larger input data size the higher the overhead, especially exceeding page size used by unified memory. The authors of [103] also tested Rodinia benchmarks modified for use with UMA and found benefits of UMA in cases when parts of data are (re)used by kernels on the GPU.

It can be noted that the UMA based approach can offer benefits over a non-UMA implementation if an application uses parts of arrays and a non-UMA approach copies whole arrays between a host and a device. On the other hand, the traditional non-UMA model, when programmed carefully and efficiently especially with overlapping communication and computations using asynchronous copy functions described in Section 4.4.5, can be potentially more efficient as the programmer has full knowledge about data and computation dependencies and can exploit these to the maximum extent.

It should also be noted that UMA is different than the Unified Virtual Addressing introduced in previous CUDA versions (compute capability 2.0+). UVA gives a unified address space for a host and a device for 64-bit applications. As described in [122], allocation of memory from a host and a device is done within UVA. Consequently, one pointer can be used to fetch data from the level of a device and a host. It is possible, though, to find out information about the pointer using a call to [123]:

```
cudaError_t cudaPointerGetAttributes(cudaPointerAttributes
*attributes, const void *pointer)
```

in order to find out where the data is stored (host or GPU and which GPU). One of the key differences between UMA and UVA is that in UMA pages with data can be migrated between a host and a device.

4.4.8 Management of GPU devices

CUDA offers several functions that allow management and usage of many GPU cards available in the host machine as well as query cards for capabilities. The following routines are typically used in such a context [123]:

```
cudaError_t cudaGetDeviceCount(int *dcount)
```

for which, upon return, `dcount` contains the number of devices that can run CUDA code,

```
cudaError_t cudaSetDevice(int devicenumber)
```

which sets the device identified with `devicenumber` as the current device. Allocating memory and subsequent launching of a kernel are performed on the current device (by default set to device 0).

```
cudaError_t cudaGetDevice(int *devicenumber)
```

which returns the current device id,

```
cudaError_t cudaChooseDevice(int *devicenumberselected,
const struct cudaDeviceProp *requestedproperties)
```

which returns (in `devicenumberselected`) a device best matching the requested `requestedproperties`,

```
cudaError_t cudaGetDeviceProperties(struct cudaDeviceProp
*properties,int devicenumber)
```

which returns useful information regarding the device identified by `devicenumber`. The most useful pieces of information are as follows (fields in structure `properties` [123]):

- string name of the device,
- size of global memory on the device,
- sizes (maximum) of shared memory per block of threads and per multiprocessor,
- numbers (maximum) of registers per block of threads and per multiprocessor,
- number of threads in a warp,
- maximum numbers of threads per block of threads and per multiprocessor,
- maximum sizes of thread block dimensions (in threads in corresponding dimensions),
- maximum sizes of grid dimensions (in blocks in corresponding dimensions),
- compute capability version,
- if it is possible to overlap computations on the device and host-GPU communication,
- number of multiprocessors as well as clock frequency,

- working mode in terms of accessing the device i.e. how many host threads will be able to call `cudaSetDevice(...)`,
- if it is possible to launch many kernels on the device from the host,
- if sharing memory space between the device and the host is enabled.

As suggested in [109] it is important to always set the current device when GPU devices are used from the level of a multithreaded host application – specifically just after new threads are created.

4.5 OPENCL

4.5.1 Programming model and application structure

OpenCL is an open standard that allows parallel programming [17, 158] targeted for heterogeneous systems, in particular including multicore CPUs and GPUs [102]. OpenCL version 2.1 was released in November 2015 while version 2.2 in May 2017. For instance, a single application can use both CPUs and GPUs installed within a server/workstation node. It should be noted, however, that such computing devices may considerably differ in performance for various types of applications which may require proper care especially in terms of load balancing. OpenCL distinguishes several key concepts, most important of which (in terms of general purpose parallel programming) include:

- Platform – defines the environment in which computations will take place. Specifically a *Host* may have one or more *Compute Device*s installed.
- Device – refers to a particular computing component on which computations will be executed. Each *Compute Device* consists of *Compute Unit*s which in turn include *Processing Element*s. Compared to the CUDA terminology, *Compute Device* corresponds to *Device* in CUDA, *Compute Unit* to *Streaming Multiprocessor* and *Processing Element* to *Streaming Processor* in CUDA.
- Context – defines a space in which kernels can be executed as well as where it is possible to manage data and synchronize processing. It includes and refers to one or more devices.
- Command queue – represents a component that allows submission of operations to a particular device for which the command queue is created. By default, commands submitted to a given queue will be performed in order. It is possible to specify out of order execution of commands though. Commands submitted to a queue include launching a kernel (processing), writing and reading memory objects (such as for initialization of input data before processing on a device and fetching results) as well as synchronization of operations.

- Program – an object representing a program that can be defined using source code. A program is created within a specified context.

- Kernel – defines a function code that can be executed in parallel by work items. The function is defined within a program that must have been created earlier.

- Memory object – typically in the context of GPGPU buffers will be used for representation of data (either input, intermediate or output) that kernel functions will operate on.

- Event – can be used for synchronization of invocations of OpenCL functions.

A parallel application in OpenCL has a structure similar to that in NVIDIA CUDA but with different names of particular components (Figure 4.7). Specifically, the application grid is called *NDRange* which consists of blocks called *work groups* which in turn include *work items*. Index spaces used for arrangement of components of lower levels may be defined in 1, 2 or 3 dimensions. Each work item can be identified using a local id within its work group and a global id within NDRange. Id dimensions correspond to dimensions defined for a work group and an NDRange respectively.

In terms of memory types defined within OpenCL, the following are distinguished:

- Global memory – all work items can read and write locations of this memory, it can also be accessed by a host thread.

- Constant memory – visibility is as for the global memory. However, while this memory can be written to by a host it can only be read from by work items.

- Local memory – accessible only to work items executing a kernel, space allocated within local memory can be shared by work items of one work group.

- Private memory – visibility such as local memory in CUDA.

It should be noted that OpenCL 2.0+ also provides Shared Virtual Memory (SVM) that enables code on the host and device sides to share presumably complex data structures [102]. In coarse-grained mode a memory buffer is shared. In fine-grained mode memory locations can be read/written concurrently if SVM atomic operations are supported or different locations can be accessed if the latter are not supported. Fine-grain buffer sharing allows buffer sharing while fine-grain system sharing allows access to the whole memory of a host.

A typical sequence of operations that are executed for parallel computations on a device in OpenCL includes:

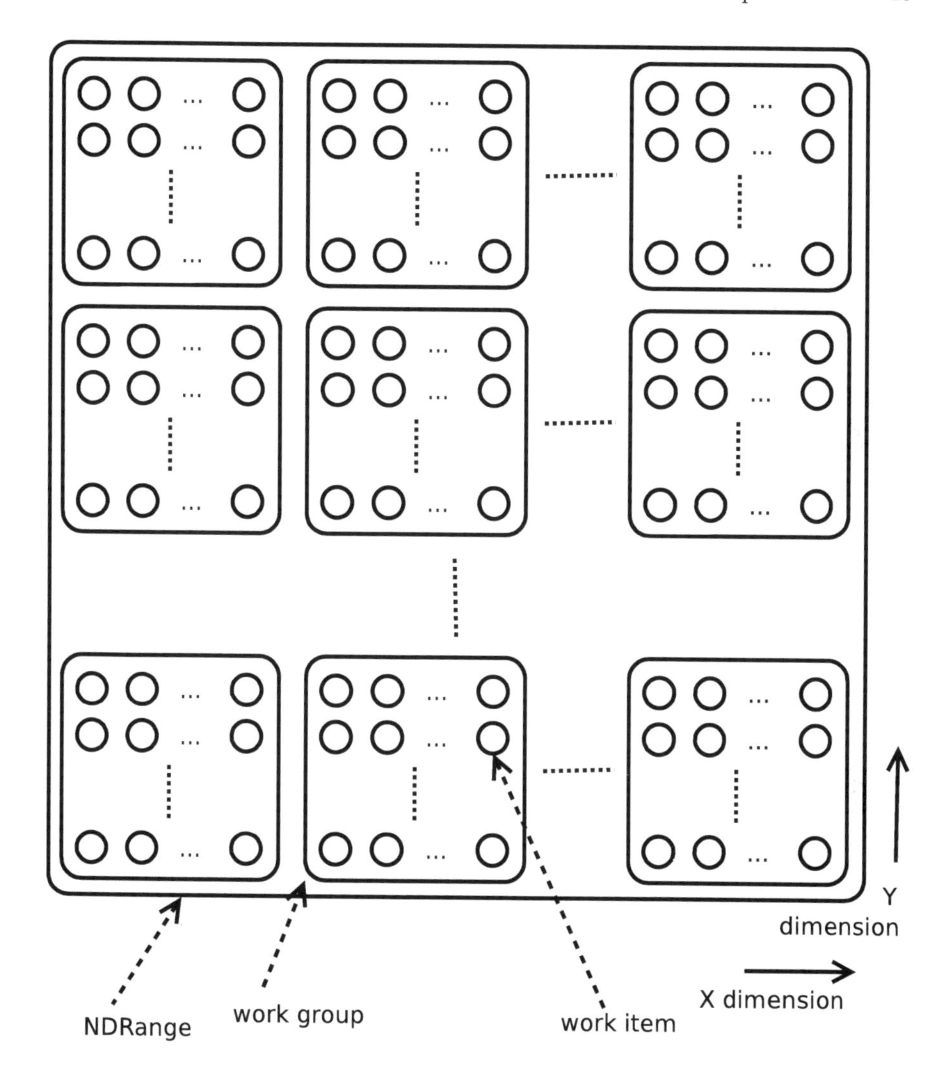

FIGURE 4.7 Architecture of an OpenCL application, a 2D example

1. Fetch a platform using a call to `clGetPlatformIDs(...)`.
2. Find out devices (either all or devices of certain types such as CPUs or GPUs within a platform using a call to `clGetDeviceIDs(...)`.
3. Create a context for a device(s) using `clCreateContext(...)`.
4. Create a command queue on a device within a context using `clCreateCommandQueueWithProperties(...)`.

5. Create a program within a context from sources given as input using `clCreateProgramWithSource(...)`.
6. Compile the program using `clBuildProgram(...)`.
7. Allocate space for input/output data that will be used by a kernel – in case of buffers this can be done using calls to `clCreateBuffer(...)` within a context.
8. Select a kernel(s) within a program using `clCreateKernel(...)`.
9. Set arguments to a kernel using function `clSetKernelArg(...)`.
10. Write input data to input buffers (if necessary) using function `clEnqueueWriteBuffer(...)` – the function queues transfer into a command queue.
11. Queue start of a kernel to a command queue using function `clEnqueueNDRangeKernel(...)`.
12. Read output data from buffers using function `clEnqueueReadBuffer (...)` – the function queues transfer from a given buffer.
13. Finish computations with `clFinish(...)` on a command queue that blocks until completion of commands submitted to the queue have been finished.

The following functions return a variable of type `cl_int` – a value equal to `CL_SUCCESS` in case of a successful call. Function [102]:

```
cl_int clGetPlatformIDs(cl_uint numberofplatformentries,
cl_platform_id *platforms,
cl_uint *numberofplatformsreturned)
```

returns available platforms – the number of platforms can be limited to `numberofplatformentries` which described the maximum number of platforms that can be stored in `platforms`. The number of returned platforms is stored in a variable pointed at by `numberofplatformsreturned`.
Function:

```
cl_int clGetPlatformInfo(cl_platform_id platform,
cl_platform_info parametername,
size_t sizeofparametervalueinbytes,
void *pointertoparametervalue,
size_t *returnedsizeofparametervalueinbytes)
```

returns information about a platform – in particular involving a name (parameter name should be `CL_PLATFORM_NAME`), version (parameter name should be `CL_PLATFORM_VERSION`) or vendor (parameter name should be `CL_PLATFORM_VENDOR`). In these cases strings are returned. Actual meaning of parameters is as follows:

- `platform` – the platform about which the call queries,
- `parametername` – name of a parameter as mentioned above,
- `pointertoparametervalue` – points to a space where a given value will be stored,
- `sizeofparametervalueinbytes` – size of space available, pointed by `pointertoparametervalue`,
- `returnedsizeofparametervalueinbytes` – returned (actual) size of value.

Furthermore, a list of available devices can be queried using the following function:

```
cl_int clGetDeviceIDs(cl_platform_id platform,
cl_device_type typeofdevice,
cl_uint numberofdeviceentries,
cl_device_id *devices,
cl_uint *numberofdevicesreturned)
```

where information about `typeofdevice` device types is queried on the given `platform`. `numberofdeviceentries` denotes how many device ids can be stored in `devices`. The number of returned devices is stored in a variable pointed at by `numberofdevicesreturned`.

Device types typically searched for include: all (`CL_DEVICE_TYPE_ALL`), CPUs (`CL_DEVICE_TYPE_CPU`), GPUs (`CL_DEVICE_TYPE_GPU`) or accelerators (`CL_DEVICE_TYPE_ACCELERATOR`).

Similarly to a platform, it is possible to find out about a particular device using function:

```
cl_int clGetDeviceInfo (cl_device_id device,
cl_device_info parametername,
size_t sizeofparametervalueinbytes,
void *pointertoparametervalue,
size_t *returnedsizeofparametervalueinbytes)
```

with the following parameters:

- `device` – the platform about which the call queries,

- `parametername` – name of a parameter referring to a particular feature of the device,
- `pointertoparametervalue` – points to a space where a given value will be stored,
- `sizeofparametervalueinbytes` – size of space available, pointed by `pointerto parametervalue`,
- `returnedsizeofparametervalueinbytes` – returned (actual) size of value.

There is a large set of information that can be queried. Selected parameters include (particular parameter values and corresponding types are available in [102]), similarly to what one can fetch in CUDA using function `cudaGetDeviceProperties(...)`:

- number of compute units,
- maximum work group and NDRange sizes,
- clock frequency for the device,
- sizes of global and local memories,
- properties of a command queue,
- maximum number of command queues,
- device and driver versions.

According to the presented initialization sequence, a thread running on a host should initialize a context which can be done with function:

```
cl_context clCreateContext(
const cl_context_properties *properties,
cl_uint numberofdevices,
const cl_device_id *devices,
void (CL_CALLBACK *pfn_notify)(const char *errorinfo,
const void *privateinfo, size_t cb,
void *callbackinputparameters),
void *callbackinputparameters,
cl_int *errorcode)
```

A context for `numberofdevices devices` is used with potential parameters specified in `properties`. A callback can be registered for runtime errors which will use the `callbackinputparameters` specified by the user.

It should be noted that OpenCL 2.0+ allows creation of host and device side queues. The latter can be used for launching kernels from the device. A command queue can be created with the following command in OpenCL 2.0+:

```
cl_command_queue clCreateCommandQueueWithProperties(
cl_context context,
cl_device_id device,
const cl_queue_properties *propertiesofqueue,
cl_int *returnederrorocode)
```

with the following parameters:

- `context` – the context in which the queue is to be created,
- `device` – a device that was initialized in `context`,
- `propertiesofqueue` – a list of desired properties for the command queue. Supported properties include `CL_QUEUE_PROPERTIES` for which a combination of possible values can be activated, including:

  `CL_QUEUE_OUT_OF_ORDER_EXEC_MODE_ENABLE` – allowing execution of commands either in order or out of order,

  `CL_QUEUE_ON_DEVICE` – defining a device queue which also implies setting `CL_QUEUE_OUT_OF_ORDER_EXEC_MODE_ENABLE`.

  According to the specification, if not defined, a host queue with in order execution is created.

  Additionally, the `CL_QUEUE_SIZE` can be used to set the queue size expressed in bytes, for device queues.
- `returnederrorocode` – returned error code in case of error, otherwise `NULL`.

Preparation of an application will typically involve specification of source code by `stringcount` `strings` each with `stringlengths` characters (`stringlengths[i]` can be 0 which requires a corresponding '\0' terminated string) including kernel(s) with function:

```
cl_program
clCreateProgramWithSource(cl_context context,
cl_uint stringcount,
const char **strings,
const size_t *stringlengths,
cl_int *returnederrorcode)
```

compiling and linking `program` for `numberofdevices` listed in `listofdevices` (can be `NULL` which builds for all devices associated with the sources) which can be done in one step with function:

```
cl_int clBuildProgram(cl_program program,
cl_uint numberofdevices,
const cl_device_id *listofdevices,
```

```
const char *options,
void (CL_CALLBACK *functionpointertonotificationfunction)(
cl_program program,
void *notificationfunctiondata),
void *notificationfunctiondata)
```

The call is blocking when `functionpointertonotificationfunction` is `NULL`, otherwise it is asynchronous. `notificationfunctiondata` is given as input to the callback function.

Creation of kernel `kernelname` can be performed using the following function for the given `program`:

```
cl_kernel clCreateKernel(cl_program program,
const char *kernel_name,
cl_int *errcode_ret)
```

and subsequent allocation of `buffersizeinbytes` memory for a buffer:

```
cl_mem clCreateBuffer(cl_context context,
cl_mem_flags bufferflags,
size_t buffersizeinbytes,
void *hostbufferpointer,
cl_int *returnederrorcode)
```

with `bufferflags` specifying in particular access mode (read-write from the point of view of the kernel by default) to the buffer such as `CL_MEM_WRITE_ONLY` (write only from the point of view of the kernel), `CL_MEM_READ_ONLY` (read only from the point of view of the kernel). A pointer to already allocated space may be passed in `hostbufferpointer` and used with `CL_MEM_USE_HOST_PTR`. Alternatively `CL_MEM_ALLOC_HOST_PTR` may be used for allocation of host accessible memory for zero copy copying. In this case `clEnqueueMapBuffer(...)` and `clEnqueueUnmapMemObject(...)` can be used on the host side for accessing the buffer. Buffers allocated in such a way may be used for computations by a kernel.

Setting kernel arguments can be done using one or more calls to function:

```
cl_int clSetKernelArg(cl_kernel kernel,
cl_uint argumentindex,
size_t argumentsize,
const void *argumentpointer)
```

where `kernel` identifies the kernel, `argumentindex` index of a given argument starting with 0, `argumentsize` denotes the size of the argument (such as the size of the buffer) and `argumentpointer` points at the input data that should be used for the argument (such as a pointer to a buffer).

It should be noted that an OpenCL program uses several qualifiers that denote corresponding elements of a program [114]:

- `__kernel` for annotating a kernel function,
- `__global` used to denote a pointer to an element or an array in global memory,
- `__constant` for annotating constant memory,
- `__local` for denoting data in local memory – can be used for a pointer in a kernel argument or within a kernel code at the function scope,
- `__private` denotes data in each work item scope – by default variables without qualifiers in functions or function arguments are of this scope.

Pointers in kernel arguments must refer to data in local, global or constant memory scopes.

There are several functions that allow to release previously allocated objects, returning `CL_SUCCESS` if finished without an error. Specifically, functions for releasing a kernel, a program, a memory object, a command queue and a context actually decrease a counter referring to particular objects. If a respective counter reaches 0 then a given object is freed. The relevant functions include:

```
cl_int clReleaseKernel(cl_kernel kernel)
```

```
cl_int clReleaseProgram(cl_program program)
```

```
cl_int clReleaseMemObject(cl_mem memoryobject)
```

```
cl_int clReleaseCommandQueue(cl_command_queue commandqueue)
```

```
cl_int clReleaseContext(cl_context context)
```

4.5.2 Coordinates and indexing

Similarly to NVIDIA CUDA, in OpenCL's kernels work items will need to find their coordinates within an NDRange. This can be accomplished with the following functions:

- `size_t get_global_id(uint idofdimension)` for getting a global index of a work item in dimension, `idofdimension` denotes identifier of a dimension (starting with 0),
- `size_t get_local_id(uint idofdimension)` for getting an index of a work item within its work group in dimension, `idofdimension` denotes identifier of a dimension (starting with 0),

– `size_t get_group_id(uint idofdimension)` for getting a global index of a work group in dimension, `idofdimension` denotes identifier of a dimension (starting with 0).

4.5.3 Queuing data reads/writes and kernel execution

Commands such as writing input buffers, spawning a kernel, reading output buffers will be submitted to a command queue. Depending on settings, execution may be performed in order or out of order.

Functions used for submitting write and read copying requests of size `sizeinbytes` from host memory `pointeronhost` to `buffer` at offset `bufferoffset` and from `buffer` at offset `bufferoffset` to host memory at `pointeronhost` are as follows:

```
cl_int clEnqueueWriteBuffer(cl_command_queue commandqueue,
cl_mem buffer,cl_bool isoperationblocking,
size_t bufferoffset,size_t sizeinbytes,
const void *pointeronhost,
cl_uint numberofeventsinwaitlist,
const cl_event *eventwaitlist,
cl_event *event)

cl_int clEnqueueReadBuffer(cl_command_queue commandqueue,
cl_mem buffer,cl_bool isoperationblocking,
size_t bufferoffset,size_t sizeinbytes,
void *pointeronhost,
cl_uint numberofeventsinwaitlist,
const cl_event *eventwaitlist,
cl_event *event)
```

Parameter `isoperationblocking` denotes whether an operation is blocking (`CL_TRUE` for yes and `CL_FALSE` for non-blocking) i.e. whether data pointed by `pointeronhost` can be reused after a function has returned (which means that the data has been read after `clEnqueueReadBuffer(...)` has completed).

It is possible to specify `numberofeventsinwaitlist` events in `eventwait list` that must be finalized before the given command is executed. On the other hand, each function returns "handler" `event` that can be used for enforcing waiting for completion of an operation is it is non-blocking as described in Section 4.5.4.

Queuing execution of `kernel` in `commandqueue` on a device is performed with function:

```
cl_int clEnqueueNDRangeKernel(cl_command_queue commandqueue,
cl_kernel kernel,
cl_uint numberofworkdimensions,
```

```
const size_t *globalworkoffsets,
const size_t *globalworksizesindimensions,
const size_t *localworksizesindimensions,
cl_uint numberofeventsinwaitlist,
const cl_event *eventwaitlist,
cl_event *event)
```

which specifies the size of an NDRange: the number of dimensions in `numberofworkdimensions`, the total numbers of work items in respective dimensions `globalworksizesindimensions`, offsets that are added to computation of a global identifier of a given work item in each dimension specified in `globalworkoffsets` (if not needed then NULL), the total numbers of work items in a work group `localworksizesindimensions`. Event related arguments are analogous to those in the previously mentioned read/write commands.

4.5.4 Synchronization functions

In OpenCL, work items of a work group can synchronize using the following barrier function:

```
void barrier(cl_mem_fence_flags flags)
```

A barrier must be executed by all work items of a work group. It is possible to specify synchronization of variables/memory fence on local or global memory if work items wish to use memory values after the barrier. These can be done using `flags` equal to `CLK_LOCAL_MEM_FENCE` or `CLK_GLOBAL_MEM_FENCE` respectively.

Furthermore, synchronization in OpenCL is possible through events. Specifically, an event is associated with function calls that submit commands to command queues. The latter return an event object. Possible states of such commands include `CL_QUEUED`, `CL_RUNNING`, `CL_COMPLETE`, or a negative integer value in case of an error.

Specifically, the following synchronization functions are available:

1. The aforementioned functions allowing submission of operations to a command queue with optional synchronization – waiting for operations associated with events.

2. Waiting on the host side for events corresponding to certain commands to complete using function:

   ```
   cl_int clWaitForEvents(cl_uint numberofevents,
   const cl_event *eventlist)
   ```

3. Submission of a marker command to a given queue – the function does not block processing but waiting for the returned `event` allows waiting later (e.g. using `clWaitForEvents(...)`) for the specified events to finalize or, if no events have been specified, to wait for all operations that have been submitted to the given queue [17] so far.

```
clEnqueueMarkerWithWaitList(cl_command_queue commandqueue,
cl_uint numberofeventsinwaitlist,
const cl_event *eventwaitlist,
cl_event *event)
```

   The function returns `CL_SUCCESS` in case of successful execution.

4. Submission of a barrier command to a given queue – commands submitted after this command will only execute after this one has completed. Waiting for the returned `event` allows waiting later for the specified events to finalize or, if no events have been specified, waiting for all operations that have been submitted to the given queue. The function itself does not block the host.

```
cl_int
clEnqueueBarrierWithWaitList(cl_command_queue commandqueue,
cl_uint numberofeventsinwaitlist,
const cl_event *eventwaitlist,
cl_event *event)
```

   The function returns `CL_SUCCESS` in case of successful execution.

5. `cl_int clFinish(cl_command_queue commandqueue)` – waits (blocks the host) until all commands submitted to `commandqueue` have been finalized. The function returns `CL_SUCCESS` in case of successful execution.

4.5.5 A sample OpenCL application

File `collatz-opencl.c` includes an example of an application that searches for the maximum number of iterations for Collatz conjecture for a range of input arguments. Listing 4.14 includes an important part of such an implementation. Each work item fetches its index with `get_global_id(0)` which is a start number for verification of the Collatz conjecture and computations of the number of iterations to reach 1. Then each work item stores its result in a separate location in global memory which is fetched to the host and searched for the maximum on the host.

Listing 4.14 Sample OpenCL implementation of a program for verification of the Collatz hypothesis

```
(...)

const char collatzkernelstring[] = "                    \
__kernel void collatzkernel                             \
    (__global long *result                              \
    )                                                   \
{                                                       \
  long myindex=get_global_id(0);                        \
  unsigned long start=myindex;                          \
  char cond=1;                                          \
  unsigned long counter=0;                              \
                                                        \
  for(;cond;counter++) {                                \
    start=(start%2)?(3*start+1):(start/2);              \
    cond=(start>1)?1:0;                                 \
  }                                                     \
                                                        \
  result[myindex]=counter;                              \
}                                                       \
";

int main(int argc,char **argv) {
  cl_platform_id platforms[1];
  cl_uint numberofplatformsreturned;
  cl_device_id devices[6];
  cl_uint numberofdevicesreturned;
  cl_context context;
  cl_int errorcode;
  cl_command_queue queue;
  cl_program program;
  size_t kernelsourcesize;
  cl_kernel kernel;
  cl_mem outputbuffer;
  long *hostbuffer;

  if (CL_SUCCESS!=clGetPlatformIDs(1,platforms,&
   numberofplatformsreturned))
    errorexit("Error getting platforms");

  if (CL_SUCCESS!=clGetDeviceIDs(platforms[0],CL_DEVICE_TYPE_GPU
    ,6,devices,&numberofdevicesreturned))
    errorexit("Error getting devices");
```

```
(...)

cl_context_properties properties[]={
  CL_CONTEXT_PLATFORM, (cl_context_properties)platforms[0],
  0};

context=clCreateContext(properties,numberofdevicesreturned,
  devices,NULL,NULL,&errorcode);
if (CL_SUCCESS!=errorcode)
  errorexit("Error creating context");

cl_queue_properties *queueproperties=NULL;
queue=clCreateCommandQueueWithProperties(context,devices[0],
                                          queueproperties,&
  errorcode);

if (CL_SUCCESS!=errorcode)
  errorexit("Error creating a command queue");

const char *kernelsource=collatzkernelstring;
kernelsourcesize=strlen(collatzkernelstring);
const char *kernelstringspointer[]={kernelsource};

program=clCreateProgramWithSource(context,1,
  kernelstringspointer,&kernelsourcesize,&errorcode);

if (CL_SUCCESS!=errorcode)
  errorexit("Error creating a program");

if (CL_SUCCESS!=clBuildProgram(program,0,NULL,"",NULL,NULL))
  errorexit("Error building program");

kernel=clCreateKernel(program,"collatzkernel",&errorcode);
if (CL_SUCCESS!=errorcode)
  errorexit("Error creating a kernel");

size_t globalworksize=1024*10000;
size_t localworksize=1024;

// allocate memory on host
if (NULL==(hostbuffer=malloc(sizeof(long)*globalworksize)))
  errorexit("Error allocating memory on host");

// create a buffer for output data
```

```
  outputbuffer=clCreateBuffer(context,CL_MEM_WRITE_ONLY,
    globalworksize*sizeof(long),NULL,&errorcode);

  if (CL_SUCCESS!=errorcode)
    errorexit("Error creating a buffer");

  if (CL_SUCCESS!=clSetKernelArg(kernel,0,sizeof(outputbuffer),&
    outputbuffer))
    errorexit("Error setting a kernel argument");

  if (CL_SUCCESS!=clEnqueueNDRangeKernel(queue,kernel,1,NULL,&
    globalworksize,&localworksize,0,NULL,NULL))
    errorexit("Error enqueuing a kernel");

  // copy results to host
  if (CL_SUCCESS!=clEnqueueReadBuffer(queue,outputbuffer,
    CL_FALSE,0,globalworksize*sizeof(long),hostbuffer,0,NULL,
    NULL))
    errorexit("Error reading results from a device");

  if (CL_SUCCESS!=clFinish(queue))
    errorexit("Error finishing computations within a queue");
  if (CL_SUCCESS!=clReleaseKernel(kernel))
    errorexit("Error releasing the kernel");
  if (CL_SUCCESS!=clReleaseProgram(program))
    errorexit("Error releasing the program");
  if (CL_SUCCESS!=clReleaseMemObject(outputbuffer))
    errorexit("Error releasing the buffer");
  if (CL_SUCCESS!=clReleaseCommandQueue(queue))
    errorexit("Error releasing the queue");
  if (CL_SUCCESS!=clReleaseContext(context))
    errorexit("Error releasing the context");

  // now find the maximum number on the host
  long result=0;
  long i;
  for(i=0;i<globalworksize;i++)
    if (hostbuffer[i]>result)
      result=hostbuffer[i];
  printf("\nResult is %ld\n",result);

  fflush(stdout);

}
```

This first code can be extended and improved with the following:

1. Finding the maximum of iterations among work items within each work group, similarly to the solution for CUDA presented in Section 4.4.4. This works by using local memory for storage of intermediate maximum iteration values in successive iterations of parallel reduction which has $O(\log(n))$ complexity where n denotes the number of elements. This step still generates one subresult per work group. The total number of subresults can still be quite large, depending on an NDRange configuration.

2. Consequently, another kernel (with one work group, assuming the number of groups is only a few times larger than the maximum number of work items in a work group) can be launched for parallel reduction of results from the previous step. In this case, the number of results from the previous step is equal to the number of work groups. These results are extended to the nearest multiple of the number of threads within a group (set to 1024 in the code) such that each work item can first browse the same number of results. Padded elements need to be initialized to 0. Specifically, each work item first finds a maximum out of `howmanyelementsperworkitem` elements. Note that in each iteration of a loop all work items collectively refer to consecutive locations in global memory, skipping 1024 elements in every iteration. Following this step work items synchronize and then collectively reduce values to a final result using the aforementioned tree based approach.

Such improved code is included in file `collatz-opencl-parallel-reduction-2-kernels.c`. An important part of such an implementation is presented in Listing 4.15.

Listing 4.15 Improved OpenCL implementation of a program for verification of the Collatz hypothesis

```
(...)
const char collatzkernelstring[] = "                    \
__kernel void collatzkernel1                           \
    (__global long *result                             \
    )                                                  \
{                                                      \
  long myindex=get_global_id(0);                       \
  int mylocalindex=get_local_id(0);                    \
  unsigned long start=myindex;                         \
  char cond=1;                                         \
  unsigned long counter=0;                             \
  __local long sresults[1024];                         \
                                                       \
  for(;cond;counter++) {                               \
```

```
    start=(start%2)?(3*start+1):(start/2);       \
    cond=(start>1)?1:0;                           \
  }                                               \
                                                  \
  sresults[mylocalindex]=counter;                 \
  barrier(CLK_LOCAL_MEM_FENCE);                   \
                                                  \
  for(counter=512;counter>0;counter/=2) {         \
    if (mylocalindex<counter)                     \
      sresults[mylocalindex]=(sresults[mylocalindex]>sresults[
    mylocalindex+counter])?sresults[mylocalindex]:sresults[
    mylocalindex+counter]; \
    barrier(CLK_LOCAL_MEM_FENCE);                 \
  }                                               \
                                                  \
  if (mylocalindex==0)                            \
    result[get_group_id(0)]=sresults[0];          \
}                                                 \
";

const char collatzkernelstring1[] = "             \
__kernel void reducekernel                        \
(__global long *data,                             \
 long howmanyelementsperworkitem,                 \
 long bufferrealelemcount                         \
 )                                                \
{                                                 \
  int mylocalindex=get_local_id(0);               \
  unsigned long counter=0;                        \
  __local long sresults[1024];                    \
                                                  \
  sresults[get_local_id(0)]=0;                    \
  if ((bufferrealelemcount+get_local_id(0))<(1024*
    howmanyelementsperworkitem)) \
    data[bufferrealelemcount+get_local_id(0)]=0;
           \
  for(counter=0;counter<howmanyelementsperworkitem;mylocalindex
    +=1024,counter++)          \
    sresults[get_local_id(0)]=(sresults[get_local_id(0)]>data[
    mylocalindex])?sresults[get_local_id(0)]:data[mylocalindex];
     \
                                                  \
  barrier(CLK_LOCAL_MEM_FENCE);                   \
                                                  \
  mylocalindex=get_local_id(0);                   \
```

```
  for(counter=512;counter>0;counter/=2) {          \
      if (mylocalindex<counter)                    \
        sresults[mylocalindex]=(sresults[mylocalindex]>sresults[
    mylocalindex+counter])?sresults[mylocalindex]:sresults[
    mylocalindex+counter]; \
      barrier(CLK_LOCAL_MEM_FENCE);                \
    }                                              \
                                                   \
  if (mylocalindex==0)                             \
    data[get_group_id(0)]=sresults[0];             \
                                                   \
}                                                  \
";

(...)

int main(int argc,char **argv) {
  cl_platform_id platforms[1];
  cl_uint numberofplatformsreturned;
  cl_device_id devices[6];
  cl_uint numberofdevicesreturned;
  cl_context context;
  cl_int errorcode;
  cl_command_queue queue;
  cl_program program;
  size_t kernelsourcesize[2];
  cl_kernel kernel,kernel1;
  cl_mem outputbuffer;
  long *hostbuffer;

  if (CL_SUCCESS!=clGetPlatformIDs(1,platforms,&
    numberofplatformsreturned))
    errorexit("Error getting platforms");


  if (CL_SUCCESS!=clGetDeviceIDs(platforms[0],CL_DEVICE_TYPE_GPU
    ,6,devices,&numberofdevicesreturned))
    errorexit("Error getting devices");

  (...)

  cl_context_properties properties[]={
    CL_CONTEXT_PLATFORM, (cl_context_properties)platforms[0],
    0};
```

```
context=clCreateContext(properties,numberofdevicesreturned,
  devices,NULL,NULL,&errorcode);
if (CL_SUCCESS!=errorcode)
  errorexit("Error creating context");

cl_queue_properties *queueproperties=NULL;
queue=clCreateCommandQueueWithProperties(context,devices[0],
                                          queueproperties,&
  errorcode);

if (CL_SUCCESS!=errorcode)
  errorexit("Error creating a command queue");

const char *kernelsource=collatzkernelstring;
kernelsourcesize[0]=strlen(collatzkernelstring);
const char *kernelsource1=collatzkernelstring1;
kernelsourcesize[1]=strlen(collatzkernelstring1);
const char *kernelstringspointer[]={kernelsource,kernelsource1
  };

program=clCreateProgramWithSource(context,2,
  kernelstringspointer,kernelsourcesize,&errorcode);
if (CL_SUCCESS!=errorcode)
  errorexit("Error creating a program");

if (CL_SUCCESS!=clBuildProgram(program,0,NULL,"",NULL,NULL))
  errorexit("Error building program");

kernel=clCreateKernel(program,"collatzkernel1",&errorcode);
if (CL_SUCCESS!=errorcode)
  errorexit("Error creating a kernel");

kernel1=clCreateKernel(program,"reducekernel",&errorcode);
if (CL_SUCCESS!=errorcode)
  errorexit("Error creating a kernel");

size_t localworksize=1024;
size_t globalworksize=1024*10000;
size_t groupcount=globalworksize/localworksize; // we assume
  that this is divisible

// now compute the smallest multiple of localworksize larger
   than groupcount
 long howmanyelementsperworkitem=groupcount/localworksize;
```

```
  if (groupcount%localworksize) howmanyelementsperworkitem+=1;
  size_t buffersize=howmanyelementsperworkitem*localworksize;

  // allocate memory on host - for just one long
  if (NULL==(hostbuffer=malloc(sizeof(long))))
    errorexit("Error allocating memory on host");

  // create a buffer for output data
  outputbuffer=clCreateBuffer(context,CL_MEM_READ_WRITE,
    buffersize*sizeof(long),NULL,&errorcode);
  if (CL_SUCCESS!=errorcode)
    errorexit("Error creating a buffer");

  if (CL_SUCCESS!=clSetKernelArg(kernel,0,sizeof(outputbuffer),&
    outputbuffer))
    errorexit("Error setting a kernel argument");

  if (CL_SUCCESS!=clEnqueueNDRangeKernel(queue,kernel,1,NULL,&
    globalworksize,&localworksize,0,NULL,NULL))
    errorexit("Error enqueuing a kernel");

// now spawn the second reduce kernel
  if (CL_SUCCESS!=clSetKernelArg(kernel1,0,sizeof(outputbuffer)
    ,&outputbuffer))
    errorexit("Error setting a kernel argument");
  if (CL_SUCCESS!=clSetKernelArg(kernel1,1,sizeof(cl_long),&
    howmanyelementsperworkitem))
    errorexit("Error setting a kernel argument");
  cl_long bufferrealelemcount=groupcount;
  if (CL_SUCCESS!=clSetKernelArg(kernel1,2,sizeof(cl_long),&
    groupcount))
    errorexit("Error setting a kernel argument");

  globalworksize=localworksize; // launch 1 work group only
  if (CL_SUCCESS!=clEnqueueNDRangeKernel(queue,kernel1,1,NULL,&
    globalworksize,&localworksize,0,NULL,NULL))
    errorexit("Error enqueuing reduce kernel");

  // copy results to host
  if (CL_SUCCESS!=clEnqueueReadBuffer(queue,outputbuffer,
    CL_FALSE,0,sizeof(long),hostbuffer,0,NULL,NULL))
    errorexit("Error reading results from a device");

  if (CL_SUCCESS!=clFinish(queue))
    errorexit("Error finishing computations within a queue1");
```

```
  if (CL_SUCCESS!=clReleaseKernel(kernel))
    errorexit("Error releasing the kernel");
  if (CL_SUCCESS!=clReleaseProgram(program))
    errorexit("Error releasing the program");
  if (CL_SUCCESS!=clReleaseMemObject(outputbuffer))
    errorexit("Error releasing the buffer");
  if (CL_SUCCESS!=clReleaseCommandQueue(queue))
    errorexit("Error releasing the queue");
  if (CL_SUCCESS!=clReleaseContext(context))
    errorexit("Error releasing the context");

  printf("\nResult is %ld\n",hostbuffer[0]);
  fflush(stdout);

}
```

The sample program can be compiled and run as follows:

```
nvcc collatz-opencl<...>.c -lOpenCL
or
gcc collatz-opencl<...>.c -lOpenCL
./a.out
```

4.6 OPENACC

4.6.1 Programming model and application structure

The OpenACC API [125] specifies ways of expressing parallelism in C and Fortran codes such that some computations can be offloaded to accelerators, typically GPUs.

While CUDA and OpenCL allow for (comparably) lower level programming with management of computing devices, explicit management of streams and queues respectively, OpenACC uses directives to point out sections of code that may include parallelism or point out specific constructs such as loops that may be parallelized.

At the level of API, OpenACC is similar to OpenMP, in the sense of using primarily directives to describe potential parallelism in a standard program.

In essence, OpenACC allows three levels of parallelism through consideration and incorporation of:

- gangs – a certain number of gangs execute,
- workers – each gang may run one or more workers,
- a certain number of vector lanes within a worker.

Similarly to OpenMP, program execution starts with a single thread. Some sections of code will employ parallelism at the aforementioned levels using constructs described next.

4.6.2 Common directives

Typical constructs used to express parallelism within a C program with OpenACC are as follows:

- Launching computations in parallel on an accelerator:

```
#pragma acc parallel
<block>
```

One or more gangs start execution of the code within the block, with one worker per gang starting execution. Similarly to OpenMP, iterations of loops within a parallel block can be processed in parallel.

Next to the `parallel` keyword, several clauses can be specified, with the most important ones being:

- `private(varlist)` – specifies that a private copy of each variable is created for every gang.
- `firstprivate(varlist)` – as in `private` but additionally each variable on the list is initialized to the value of the thread before the construct.
- `reduction(operator:varlist)` – similarly to OpenMP, private copies for each gang are created and their results are merged into a final result and the latter can be used after the construct has been terminated.
- `if(condition)` – if the condition is true then the following block will be executed on a device, otherwise by the original thread.

Furthermore, configuration can be controlled using optional clauses such as:

- `num_gangs(number)` – specifies the number of gangs that are requested to execute the block (in reality it may be lower).
- `num_workers(number)` – specifies the number of workers within a gang when multiple workers are used for execution (such as loop parallelization among workers).
- `vector_length(number)` – specifies the number of vector lanes used when parallelizing loops such that loops are assigned to vector lanes first (see below).

By default, similarly to OpenMP, an implicit barrier is enforced at the end of the parallel region. If not needed, an `async` clause can be used.

- Marking a code region which may contain code that can be run as a sequence of kernels:

```
#pragma acc kernels
<block>
```

Parallelization of configurations (number of gangs, workers and vectors) for kernels might differ. Clauses `num_gangs(number)`, `num_workers (number)` and `vector_length(number)` can be used similarly as in the case of the `parallel` construct.

– Marking a for loop for which parallel execution might be applied:

```
#pragma acc loop
<loop>
```

The `gang` parameter specifies distribution of loop iterations for parallel execution among gangs. The `static:<sizeofchunk>` keyword specifies a chunk size and round robin assignment of loop iterations. The runtime system may select the chunk size if `*` is used for the chunk size.

Furthermore, the `worker` keyword instructs distribution of loop iterations among workers in a gang.

The `vector` keyword instructs distribution of loop iterations among vector lanes.

The `tile` keyword instructs conversion of a loop in a loop nest to two loops to exploit locality during computations [104].

`private(varlist)` of a `loop` construct makes the specified variables private at the used parallelization level for the loop. The `reduction(operator:varlist)`, similarly to OpenMP, will create private copies of the specified variables and merge values into a final value using the given operator.

4.6.3 Data management

OpenACC considers two reference counts – structured incremented when entering a region and decremented upon exit and dynamic incremented for `enter data` and decremented for `exit data` clauses as described next.

There are several clauses referring to data management that may be used with `kernels` or `parallel` constructs.

These include, in particular:

- `create(varlist)` – allocates memory on a device if it does not exist, a reference count is increased. When exiting a region, the reference count is decreased.

- `copy(varlist)` – allocates memory on a device if it does not exist, a reference count is increased and data is copied to the device. When exiting a region, the reference count is decreased. If both reference counts are equal to 0 then data is copied from the device to the host and space on the device is freed.
- `copyin(varlist)` – similarly to `copy` but data is not sent back to the host upon exiting a region.
- `copyout(varlist)` – similarly to `copy` but data is not sent to the device when entering a region.
- `present(varlist)` – specifies that the data is present on a device, the reference count is incremented when entering a region, decremented upon exiting.

Additionally, it is possible to use the following constructs:

– `data` construct that defines data management within a following block with the following syntax:

```
#pragma acc data <clause>
<block>
```

where, in particular, the following clauses are allowed with meaning described above: `create`, `copy`, `copyin`, `copyout` and `present`.

– `enter data` for definition of data on a device until `exit data` is stated or a program exits. Specifically, the syntax is as follows:

```
#pragma acc enter data <clause>
```

where a clause can be, in particular, `create`, `copyin`, `if`.

Similarly, the following can be used at exit:

```
#pragma acc exit data <clause>
```

where a clause can be, in particular, `copyout`, `delete`, `if`.

OpenACC allows expression of operations to be atomic to avoid issues resulted from races. Specifically, the syntax is as follows:

```
#pragma acc atomic <clause>
<expression>
```

where a clause can be one of the following referring to a variable being accessed: `read`, `write`, `update` (increments, decrements, binary operations), `capture` (update with a read).

4.6.4 A sample OpenACC application

An OpenACC version of the OpenMP based application presented in Section 4.2.6 is included in file lnx_OpenACC_0.c. An important part of such an implementation is shown in Listing 4.16.

Listing 4.16 Basic OpenACC implementation of computation of $\ln(x)$

```
  (...)

  int i;
  double x;
  long maxelemcount=10000000;
  double sum=0; // final value
  long power; // acts as a counter

  (...)

  // read the argument
  x=...

  // read the second argument
  maxelemcount=...

  sum=0;
  // use the OpenACC parallel loop construct

#pragma acc parallel loop reduction(+:sum)
  for (power=1;power<maxelemcount;power++)
    sum=sum+pow(((x-1)/x),(double)power)/power;

  printf("\nResult is %f\n",sum);

  (...)
```

In this particular case, the accULL [138, 139] version 0.4 alpha software was used for compilation and running the aforementioned example in a Linux based system.

There are several OpenACC compilers including the OpenACC Toolkit available from NVIDIA [121] with a PGI Accelerator Fortran/C Compiler, support for OpenACC in GCC [2] etc.

4.6.5 Asynchronous processing and synchronization

OpenACC allows to use an async clause [125], especially in the parallel and kernels constructs but also, in particular, in enter data and exit data constructs. Without it execution of corresponding blocks would block the initiating thread until execution finishes. With `async(intargument)` a block would be able to execute asynchronously with respect to the initiating thread. The argument is used to identify a queue on a device to which execution will be submitted, if the device supports multiple queues.

Correspondingly, a `wait` clause can be used in the `parallel` and `kernels` constructs but also, in particular, in `enter data` and `exit data` constructs. By default, execution of corresponding blocks starts execution at once. The `wait` clause with an argument such as used in `async` clauses, can be used to force waiting with execution until corresponding computations (identified by the argument) complete.

It is also possible to use a `wait` directive with the following syntax:

```
#pragma acc wait [intargument]
```

which allows waiting for completion of computations in a given queue or all operations submitted from the thread. It is also possible to use an async clause at the end of this directive which allows waiting with submission to the async queue until the wait argument queue completes.

4.6.6 Device management

An OpenACC runtime library allows management and use of several devices. Among many functions of the library, the following ones can be used for this purpose:

```
void acc_set_device_type(acc_device_t devicetype)
```

which sets the type of a device to be used,

```
void acc_set_device_num(int devicenumber,
acc_device_t devicetype)
```

which sets the active device (number) of the given type,

```
int acc_get_num_devices(acc_device_t devicetype)
```

which obtains the number of available devices of the given type.

4.7 SELECTED HYBRID APPROACHES

While the aforementioned APIs are often used individually for parallel programming, it is possible and often beneficial to combine various APIs for best performance and/or easy mapping of a parallel application to a high performance computing system, especially if many levels of parallelism are involved.

The following sections include selected combinations of APIs allowing parallelization at various levels.

More hybrid examples implementing master-slave, geometric SPMD and divide and conquer paradigms described in Chapter 3 are included in Sections 5.1, 5.2 and 5.3.

4.7.1 MPI+Pthreads

File `lnx_MPI+Pthreads.c` includes a parallel code that uses MPI and the Pthreads API to implement the problem stated in Section 4.1.13, first to start processes working in parallel and the latter to start threads within processes. Threads dedicated for computations are started with a call to `pthread_create(...)` and execute function `void *Calculate(void *args)`. Within this function, that can be implemented as follows:

```
void *Calculate(void *args) {

  int start=*((int *)args); // start from this number
  double partialsum=0; // partial sum computed by each process
  double mult_coeff; // by how much multiply in every iteration
  double prev; // temporary variable
  long power; // acts as a counter
  long count; // the number of elements

  // each process performs computations on its part

  mult_coeff=(x-1)/x;
  count=maxelemcount/totalthreads;
  // now compute my own partial sum in a loop
  power=(myrank*threadnum+start)*count+1;
  prev=pow(((x-1)/x),power);
  for(;count>0;power++,count--) {
    partialsum+=prev/power;
    prev*=mult_coeff;
  }
  threadResults[start]=partialsum;
}
```

Each thread initializes the power it will start with by fetching the value of its process rank as well as its thread id passed as an argument to function `Calculate(...)`. After each thread has computed its partial result it writes the result to a global array. Each of these values is added after a call to `pthread_join(...)` from the main thread for which the key part of the code is shown in Listing 4.17. Then all results from processes are added with a call to `MPI_Reduce(...)`.
The sample program can be compiled and run as follows:

```
mpicc <flags> lnx_MPI+Pthreads.c -lm
mpirun --bind-to none -np 2 ./a.out 300236771.4223 10000000000
```

Listing 4.17 Parallel implementation of computing $\ln(x)$ using combined MPI and Pthreads

```
(...)
double totalprocesssum=0;
double totalsum=0;
(...)
int main(int argc, char **argv) {

  int i;
  int threadsupport;
  void *threadstatus;

  // Initialize MPI with support for multithreading
  MPI_Init_thread(&argc, &argv,MPI_THREAD_FUNNELED,&
    threadsupport);
  if (threadsupport<MPI_THREAD_FUNNELED) {
    printf("\nThe implementation does not support
    MPI_THREAD_FUNNELED, it supports level %d\n",threadsupport);
    MPI_Finalize();
    return -1;
  }

  // check if an argument was given to the program
  (...)

  // read the argument
  x=...

  // read the second argument if provided
  if (argc>=3)
    maxelemcount=...

  // read the number of threads if provided
  if (argc>=4)
    threadnum=...

  (...)

  // find out the rank of this process
  MPI_Comm_rank(MPI_COMM_WORLD, &myrank);
  // find out the number of processes
```

```
  MPI_Comm_size(MPI_COMM_WORLD, &proccount);

  if (maxelemcount<proccount) {
    if (!myrank)
      printf("Maxelemcount smaller than the number of processes
    - try again.");
    MPI_Finalize();
    return -1;
  }

  // now start the threads in each process
  // define the thread as joinable
  pthread_attr_init(&attr);
  pthread_attr_setdetachstate(&attr, PTHREAD_CREATE_JOINABLE);

  // initialize the totalthreads value
  totalthreads=proccount*threadnum;

  for (i=0;i<threadnum;i++) {
    // initialize the start value
    startValue[i]=i;
    // launch a thread for calculations
    pthread_create(&thread[i], &attr, Calculate, (void *)(&(
    startValue[i])));
  }
  // now synchronize the threads
  // and add results from all the threads in a process
  for (i=0;i<threadnum;i++) {
    pthread_join(thread[i], &threadstatus);
    totalprocesssum+=threadResults[i];
  }
  // now merge results from processes to rank 0
  MPI_Reduce(&totalprocesssum,&totalsum,1,
             MPI_DOUBLE,MPI_SUM,0,
             MPI_COMM_WORLD);
  if (!myrank)
    printf("Result=%f\n",totalsum);
  pthread_attr_destroy(&attr);
  // Shut down MPI
  MPI_Finalize();
  return 0;
}
```

Figure 4.8 presents execution times for various numbers of processes and computing threads per process started with Pthreads run on a workstation with 2 x Intel Xeon E5-2620v4 and 128 GB RAM. This type of application does not benefit much from HyperThreading. In such an environment it is generally better to use more threads per process than a larger number of processes with fewer threads due to lower overhead for creation and switching between threads compared to processes. Consequently, in a cluster environment it is recommended to use MPI for internode communication with parallelization within a node using threads. Results are best times out of 3 runs for each configuration.

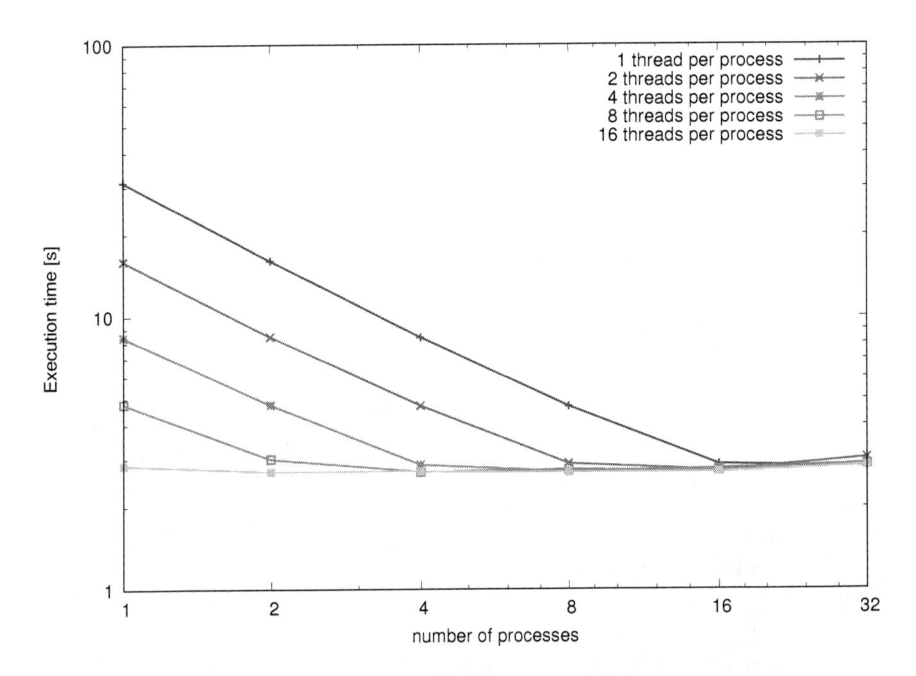

FIGURE 4.8 Execution time of the testbed MPI+Pthreads application run on a workstation with 2 x Intel Xeon E5-2620v4, 128 GB RAM, 16 physical cores, 32 logical processors, parameters: 300236771.4223 20000000000

It should be noted, that according to the OpenMPI manual, `mpirun` binds processes at the start depending on the number of processes. If threads are used (such as in this case), either no binding is suggested (as imposed by `--bind-to none` as in the example above) or binding to many cores. In the experiments, in general `--bind-to none` gave better results for 1 and 2 processes but better results were obtained without this option for larger numbers of processes. Best times are presented.

4.7.2 MPI+OpenMP

Extending the previous implementations, the code included in file `lnx_MPI+OpenMP.c` combines MPI and OpenMP in one application for both parallel processing among processes and within each process using threads. Listing 4.18 lists key parts of such an implementation. An OpenMP `#pragma omp parallel` directive is used for parallel processing using threads in each process with summing thread results with the `reduction (+:partialsum)` clause.

Listing 4.18 Parallel implementation of computing $\ln(x)$ using combined MPI and OpenMP

```
int i;
int threadsupport;
long maxelemcount=1000000000;
int totalthreads; // how many threads in total
int myrank,proccount; // rank of a given process and the
  number of processes
double x; // argument
double partialsum=0; // partial sum computed by each process
double totalsum=0;
double mult_coeff; // by how much multiply in every iteration
double prev; // temporary variable
long power; // acts as a counter
int numberofthreads;
long count; // the number of elements

// Initialize MPI with support for multithreading

MPI_Init_thread(&argc, &argv,MPI_THREAD_FUNNELED,&
  threadsupport);

if (threadsupport<MPI_THREAD_FUNNELED) {
  printf("\nThe implementation does not support
  MPI_THREAD_FUNNELED, it supports level %d\n",threadsupport);
  MPI_Finalize();
  return -1;
}

// check if an argument was given to the program
if (argc<3) {
  if (!myrank)
    printf("\nSyntax: lnx <x> <nelem>");
  MPI_Finalize();
  return -1;
```

```
  }

  // read the argument
  x=...

  // read the second argument if provided
  if (argc>=3)
    maxelemcount=...

  // read the number of threads if provided
  if (argc>=4)
    threadnum=...

  // find out the rank of this process
  MPI_Comm_rank(MPI_COMM_WORLD, &myrank);
  // find out the number of processes
  MPI_Comm_size(MPI_COMM_WORLD, &proccount);

  if (maxelemcount<proccount) {
    if (!myrank)
      printf("Maxelemcount smaller than the number of processes
    - try again.");
    MPI_Finalize();
    return -1;
  }

  // now start the threads and process data within threads

#pragma omp parallel shared(proccount,myrank) private(prev,power
    ,totalthreads,numberofthreads,mult_coeff,count) reduction
    (+:partialsum) num_threads(threadnum)
  {
    // check how many threads actually created
    numberofthreads=omp_get_num_threads();
    totalthreads=proccount*numberofthreads;
    mult_coeff=(x-1)/x;
    count=maxelemcount/totalthreads;
    // now compute my own partial sum in a loop
    power=(myrank*numberofthreads+omp_get_thread_num())*count+1;
    prev=pow(((x-1)/x),power);
    for(;count>0;power++,count--) {
      partialsum+=prev/power;
      prev*=mult_coeff;
    }
  }
```

```
// now merge results from processes to rank 0
MPI_Reduce(&partialsum,&totalsum,1,
           MPI_DOUBLE,MPI_SUM,0,
           MPI_COMM_WORLD);
if (!myrank)
  printf("Result=%f\n",totalsum);
// Shut down MPI
MPI_Finalize();
```

The code can be compiled and run as follows:

```
mpicc -fopenmp <flags> lnx_MPI+OpenMP.c -lm
mpirun --bind-to none -np 2 ./a.out 300236771.4223 20000000000
```

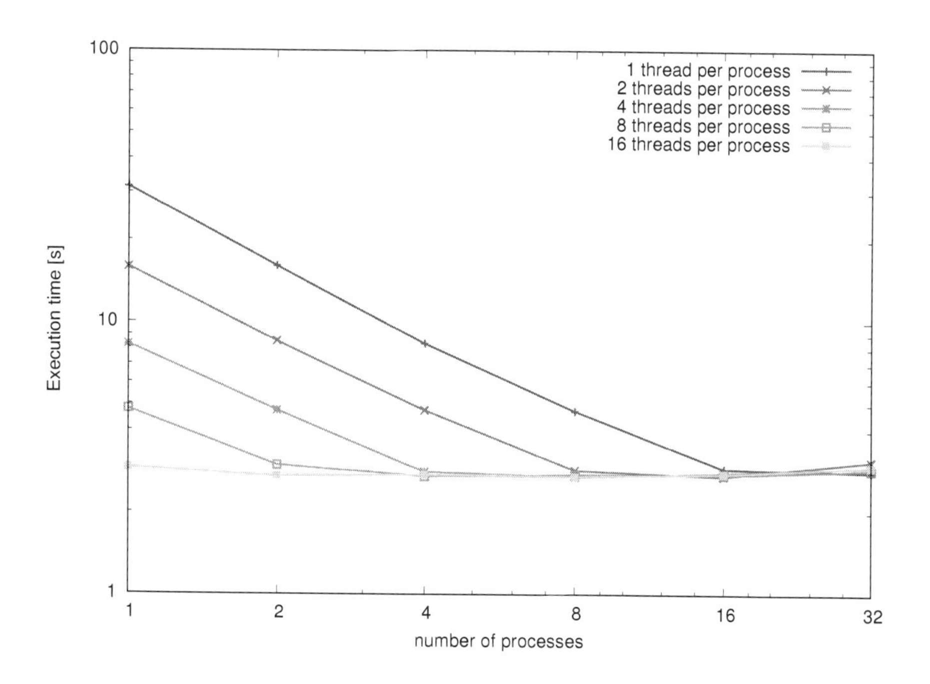

FIGURE 4.9 Execution time of the testbed MPI+OpenMP application run on a workstation with 2 x Intel Xeon E5-2620v4, 128 GB RAM, 16 physical cores, 32 logical processors, parameters: 300236771.4223 20000000000

Figure 4.9 presents results of the testbed MPI+OpenMP application run on a workstation with 2 x Intel Xeon E5-2620v4, 128 GB RAM. Best times out of three runs for each configuration are presented.

In the experiments, `--bind-to none` gave better results for 1 and 2 processes but generally better results were obtained without this option for larger numbers of processes. Best times are presented.

4.7.3 MPI+CUDA

The parallel application for verification of the Collatz hypothesis, presented in Section 4.4.4, can be extended to be used in a cluster environment, with potentially more than one GPU per node. The goal of the code is not only to verify the hypothesis (in which case computations for each considered starting number should eventually terminate) but also to compute the maximum number of steps for the starting number to reach 1 with the operations listed in the Collatz hypothesis.

The structure of such an application is as follows:

- `collatz-MPI+CUDA.c` file includes MPI initialization, fetching the number of processes, a given process rank, invocation of function `launchcomputekernel(...)` and finalization of MPI. For the sake of simplicity of implementation, each process manages computations on one GPU. If there are more GPUs per node, the number of processes equal to the number of GPUs can be spawned – it is assumed that processes are launched in batches (successive ranks in each batch) on successive nodes. Obviously, such an implementation can be further extended for incorporation of computations on CPU cores as well. An important part of such an implementation is shown in Listing 4.19.

Listing 4.19 MPI and CUDA implementation of verification of the Collatz hypothesis – host code

```
(...)
int launchcomputekernel(int deviceno,long start,long end,
    long *result,int myrank,int processcount);

int main(int argc,char **argv) {
  long start,end;
  int processcount,myrank;
  int mygpuid;
  int gpuspernode;
  long result,finalresult;

  MPI_Init(&argc, &argv);

  (...)

  start=atol(argv[1]);
  end=atol(argv[2]);
```

```
  gpuspernode=atoi(argv[3]);

  // find out the total number of processes
  MPI_Comm_size(MPI_COMM_WORLD,&processcount);
  // and my rank
  MPI_Comm_rank(MPI_COMM_WORLD,&myrank);

  // we assume that each process is responsible for a
    different GPU
  // now launch computations on "my" GPU

  // assuming divisible numbers

  if (!launchcomputekernel(myrank%gpuspernode,start,end,&
    result,myrank,processcount))
    errorexit("\nError on the .cu side");

  // reduce values from all processes
  MPI_Reduce(&result,&finalresult,1,MPI_LONG,MPI_MAX,0,
    MPI_COMM_WORLD);

  if (!myrank)
    printf("\nFinal result is %ld\n",finalresult);

  MPI_Finalize();
}
```

- `CUDAkernel.cu` file that includes implementation of function `launchcomputekernel(...)` which does the following:
  1. Selects an active GPU.
  2. Allocates space for the maximum number of iterations per each block.
  3. Spawns kernel function `checkCollatz(...)`. It should be noted that various MPI processes get starting numbers that are interleaved. The goal of this partitioning is to make sure that starting numbers of similar absolute values are distributed among various processes and consequently various GPUs. Additionally, after each thread within a block has computed the number of iterations for its starting number, the maximum number per block is computed using reduction in shared memory.
  4. Copies results back from the global memory to the host memory.
  5. The maximum number is found on the host.

Listing 4.20 presents key lines of relevant implementation of this part of the code – kernel and its invocation.

Listing 4.20 MPI and CUDA implementation of verification of the Collatz hypothesis – kernel and kernel invocation

```
(...)
__global__
void checkCollatz(long *result,long beginning,int myrank,
    int processcount) {
  long my_index=beginning+myrank+processcount*(blockIdx.x*
    blockDim.x+threadIdx.x);
  unsigned long start=my_index;
  char cond=1;
  unsigned long counter=0;
  __shared__ long sresults[1024];

  for(;cond;counter++) {
    start=(start%2)?(3*start+1):(start/2);
    cond=(start>1)?1:0;
  }
  sresults[threadIdx.x]=counter;
  __syncthreads();

  // now computations of a max within a thread block can
    take place
  for(counter=512;counter>0;counter/=2) {
    if (threadIdx.x<counter)
      sresults[threadIdx.x]=(sresults[threadIdx.x]>sresults
    [threadIdx.x+counter])?sresults[threadIdx.x]:sresults[
    threadIdx.x+counter];
    __syncthreads();
  }

  // store the block maximum in a global memory
  if (threadIdx.x==0)
    result[blockIdx.x]=sresults[0];
}

extern "C" int launchcomputekernel(int deviceno,long start,
    long end,long *result,int myrank,int processcount) {
  int threadsinblock=1024;
  int blocksingrid=(end-start+1)/(1024*processcount); //
    assuming divisible

  long size=blocksingrid*sizeof(long);
```

```
long *hresults=(long *)malloc(size);
if (!hresults)
  return 0;

if (cudaSuccess!=cudaSetDevice(deviceno)) { // each
  process sets an active GPU on the node it will manage
  printf("\nError setting a CUDA device. Check the number
   of devices per node.\n");
  return 0;
}

long *dresults=NULL;
if (cudaSuccess!=cudaMalloc((void **)&dresults,size)) {
  printf("\nError allocating memory on the GPU");
  return 0;
}

// start computations on the GPU
checkCollatz<<<blocksingrid,threadsinblock>>>(dresults,
  start,myrank,processcount);
if (cudaSuccess!=cudaGetLastError()) {
  printf("\nError during kernel launch");
  return 0;
}

if (cudaSuccess!=cudaMemcpy(hresults,dresults,size,
  cudaMemcpyDeviceToHost)) {
   printf("\nError copying results");
   return 0;
}
cudaDeviceSynchronize();

// now find the maximum number on the host
*result=0;
for(int i=0;i<blocksingrid;i++)
  if (hresults[i]>*result)
    *result=hresults[i];

// release resources
free(hresults);
if (cudaSuccess!=cudaFree(dresults)) {
  printf("\nError when deallocating space on the GPU");
  return 0;
}
return 1;
```

```
}
```

Such code organization means that the following steps can be used in order to compile and link in order to obtain a final executable file:

```
nvcc -arch=sm_35 -c CUDAkernel.cu -o CUDAkernel.o
mpicc collatz-MPI+CUDA.c CUDAkernel.o -lcudart -lstdc++
```

which can be invoked as follows, as an example:

```
mpirun -np 2 ./a.out 1 102400000 2
```

The first two arguments denote the start and the end of the range to check and the last argument denotes the number of GPUs per node that the application will use.

Results of running this application on a workstation with 2 x Intel Xeon E5-2620v4 and 128 GB RAM and two NVIDIA GTX 1070 cards are presented in Table 4.4.

TABLE 4.4 Execution times [s] for the MPI+CUDA code

number of processes of MPI application	2 x Intel Xeon E5-2620v4 + 2 x NVIDIA GTX 1070
1	28.067
2	15.867

CHAPTER 5

Programming parallel paradigms using selected APIs

CONTENTS

5.1 MASTER-SLAVE

Implementations of the master-slave paradigm may differ in how the following actions are executed:

- data partitioning – whether it is performed statically before computations or dynamically by a master,
- merging results from slaves – whether it is performed dynamically after each result has been received by the master or after all results have been gathered.

Furthermore, the code can be simplified in such a way that the active role of the master is eliminated by transferring master's responsibilities to slaves that fetch successive data packets, compute, store results and repeat this process. In this case, a synchronization mechanism among slaves needs to be adopted.

5.1.1 MPI

This section proposes a code (available in file `master-slave-MPI.c`) of a basic master-slave application implemented with C and MPI. The application is to perform numerical integration of a given function (defined within C function `fcpu(...)`) over a given range. The function can be defined, as an example, as follows:

```
double fcpu(double x) {
  return 1.0/(1.0+x);
}
```

For the sake of generalization of the code, several elements realizing the master-slave concept are distinguished:

1. Definition of a data type – `functionparameters_t` that encapsulates a certain number of elements and corresponding results:

   ```
   typedef struct {
     long elemcount;
     double *elements;
     double *result;
   } functionparameters_t;
   ```

2. Function `generatedata(...)` that generates a certain number of data packets (input data) that will be processed within the application. It returns a pointer to data packets as well as the number of data packets generated. Input parameter `datapacketmultiplier`, which is also an input parameter to the program, can be used for imposing a larger or a smaller number of data packets.
3. Function `resultmergecpufunction(...)` that merges input elements into a final result. The implementation of the function itself is bound to

a specific problem. For instance, it may just add subresults – as in the proposed code.

4. Function `slavecpu(...)` that processes a single data packet.

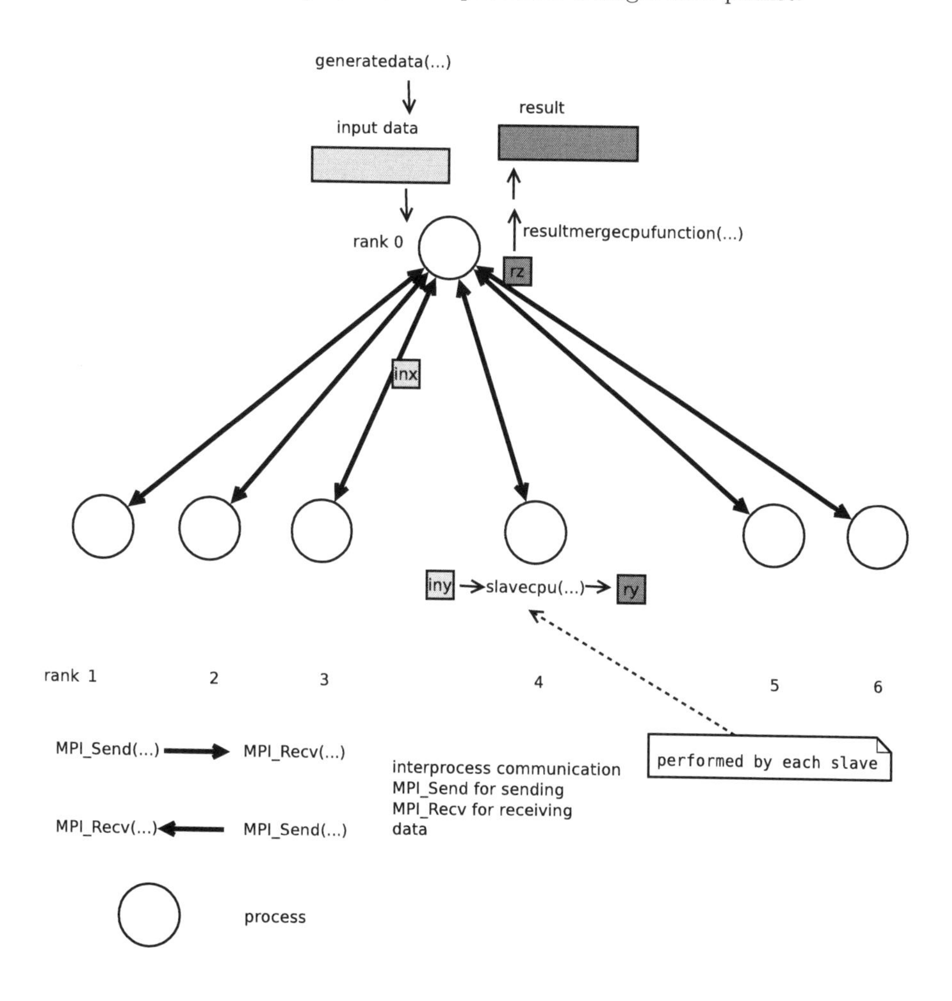

FIGURE 5.1 Master-slave scheme with MPI

The master-slave scheme is shown in Figure 5.1 and discussed in more detail below. In the `main(...)` function the master (process with rank 0) does the following (Listing 5.1 presents key code of the master process):

1. Generates input data packets by partitioning the input range into a number of subranges.

2. Distributes one subrange per slave (any process other than the master becomes a computing process – slave).

3. Waits for completion of processing of any data packet (from MPI_ANY_SOURCE), merges the result and responds back with a new data packet as long as a new packet is still available.

4. Gathers all remaining results.

5. Sends a termination message to all slaves.

Listing 5.1 Basic master-slave application using MPI – master's key code

```
long packetcounter=0;
// distribute initial data packets to slave processes
for(i=1;i<processcount;i++,packetcounter++)
  MPI_Send(&(data[2*packetcounter]),2,MPI_DOUBLE,i,DATA_PACKET,
    MPI_COMM_WORLD);

// now wait for results and respond to incoming result messages
// do it individually, in general data packets may take more or
    less time
// to process (e.g. due to various CPUs on various nodes)
do {
  MPI_Recv(&temporaryresult,1,MPI_DOUBLE,MPI_ANY_SOURCE,RESULT,
    MPI_COMM_WORLD,&mpistatus);
  // merge result
  resultmergecpufunction(&args);          // result in variable
    result

  // check who sent the result and send another data packet to
    that process
  MPI_Send(&(data[2*packetcounter]),2,MPI_DOUBLE,mpistatus.
    MPI_SOURCE,DATA_PACKET,MPI_COMM_WORLD);
  packetcounter++;

} while (packetcounter<packetcount); // do it as long as there
    are data packets to send

// now wait for pending results
for(i=1;i<processcount;i++) {
  MPI_Recv(&temporaryresult,1,MPI_DOUBLE,MPI_ANY_SOURCE,RESULT,
    MPI_COMM_WORLD,&mpistatus);
  resultmergecpufunction(&args);          // result in variable
    result
}

// and send a termination message
for(i=1;i<processcount;i++)
```

```
  MPI_Send(NULL,0,MPI_DOUBLE,i,FINISH_COMPUTATIONS,
    MPI_COMM_WORLD);
// display the result
printf("\nResult is %f\n",result);
```

Each slave acts in a loop and does the following (Listing 5.2 presents key code of the slave process):

1. Checks for an incoming message in order to make sure whether a new data packet is being sent or whether it is a termination message – this is done with function `MPI_Probe(...)`.
2. Receives a message if a data packet is incoming.
3. Processes the data packet.
4. Sends a result back.

Listing 5.2 Basic master-slave application using MPI – slave's key code

```
double data[2];
do {
  MPI_Probe(0,MPI_ANY_TAG,MPI_COMM_WORLD,&mpistatus);
  if (mpistatus.MPI_TAG==DATA_PACKET) {
    MPI_Recv(data,2,MPI_DOUBLE,0,DATA_PACKET,MPI_COMM_WORLD,&
    mpistatus);
    // spawn computations in parallel using OpenMP
    step=(data[1]-data[0])/(40*DATA_PACKET_PARTITION_COUNT); //
    equivalent to other implementations

    slavecpu(&temporaryresult,data[0],data[1],step);
    // send the result back
    MPI_Send(&temporaryresult,1,MPI_DOUBLE,0,RESULT,
    MPI_COMM_WORLD);
  }
} while (mpistatus.MPI_TAG!=FINISH_COMPUTATIONS);
```

The application can be compiled as follows:

```
mpicc <flags> master-slave-MPI.c
```

and run in the following way, with 1 master and 16 slave processes

```
mpirun -np 17 ./a.out 1
```

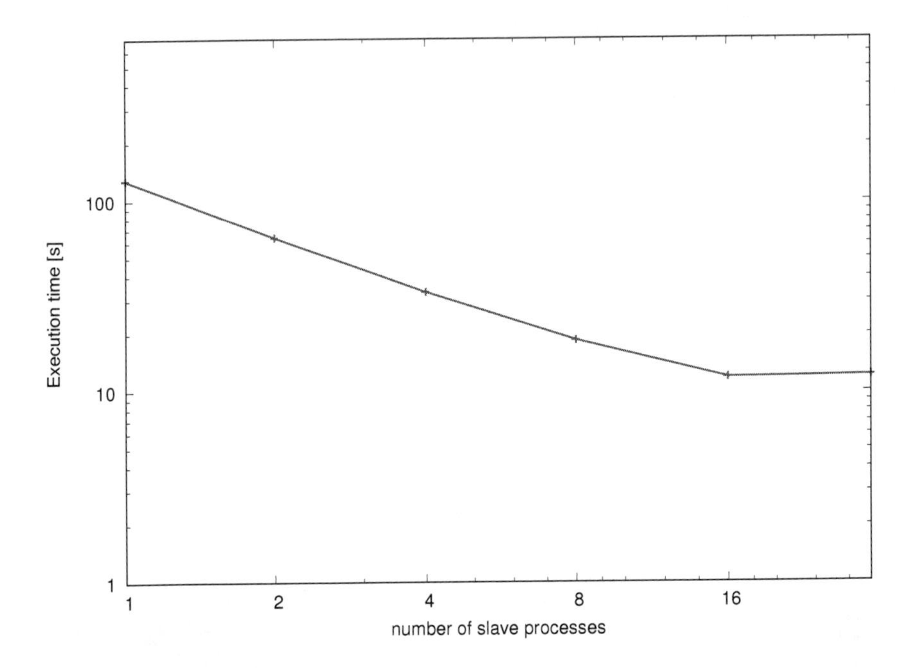

FIGURE 5.2 Execution time of the testbed MPI master-slave application run on a workstation with 2 x Intel Xeon E5-2620v4, 128 GB RAM, 16 physical cores, 32 logical processors

Execution times obtained on a workstation with 2 x Intel Xeon E5-2620v4 and 128 GB RAM are shown in Figure 5.2. For each configuration, the best out of three runs are reported.

It should be noted that one of the first potential optimizations of the code would be to allow processing of a data packet by a slave and master-slave communication at the same time. Implementation of such a technique is discussed in Section 6.1.

5.1.2 OpenMP

OpenMP allows implementation of master-slave type parallel programming in several ways. Specifically, the traditional master-slave requires the following:

1. Distinguishing and identification of a master thread and slave threads. This can be done in a few ways:

   - Using the `sections` directive and running master and slave codes in various `section` blocks.

- Using function `omp_get_thread_num()` within a parallel block to fetch thread ids and running a master code for e.g. thread with id 0 and a slave code for threads with other ids.
- Using the `#pragma omp master` directive to specify code that will be executed by the master thread in a team of threads.

2. Communication and synchronization when fetching input data and merging results. In this respect OpenMP does provide some possibilities:

   - Synchronization using variables that can be accessed by both master and slave threads, as demonstrated in [165]. In this case, such variables can act as semaphores. After a thread has updated such a synchronization variable, the thread flushes it with directive `#pragma omp flush (variable)` so that threads have a consistent view of memory, as discussed in Section 4.2.4.
   - Synchronization through locks available in the OpenMP API, specifically using functions:
     - `omp_set_lock(omp_lock_t *lock)` – waiting until the given lock is available and proceeding with execution,
     - `omp_unset_lock(omp_lock_t *lock)` – makes the lock available.

The approach shown in Figure 5.3 adopts the very traditional implementation of the master-slave in which the master thread generates input at runtime and also merges results at runtime. A disadvantage of this approach is active waiting involved. Locks for input and output are used. This version is improved further on. Its assumptions are as follows:

1. There are input and output buffers that can store `BUFFERSIZE` elements each. Types `t_input` and `t_output` are used respectively.
2. One thread acts as a master that initially generates input data into the input buffer. The thread checks periodically whether the data from the input buffer has been fetched by slave threads and processed. If so, it fills in the input buffer again.
3. Each slave thread fetches an index of a new input data chunk, processes it and stores a result in the output buffer. Then it repeats the procedure.
4. The master checks whether all data in the input data buffer has been processed. If so, it merges results into the final result and generates new input data into the input buffer.

It is possible to modify and improve such an approach (presented in file `master-slave-OpenMP-0.c`) in such a way that no dedicated master thread is distinguished. Instead, the following conceptual modifications, presented in Figure 5.4, can be introduced:

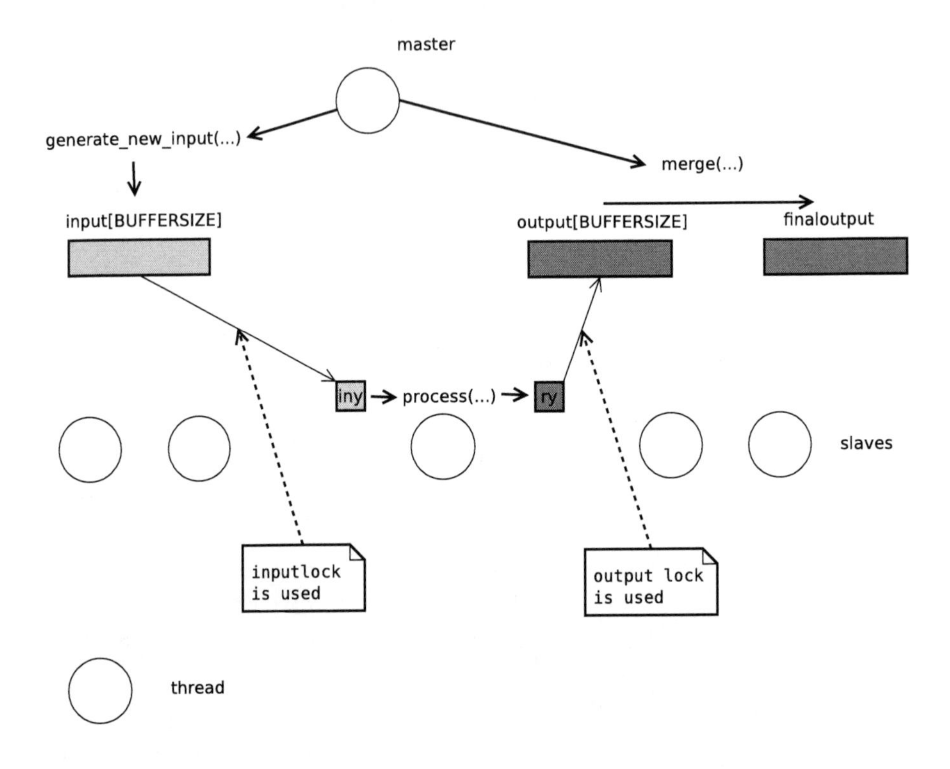

FIGURE 5.3 Master-slave scheme with OpenMP – initial version

1. All threads are in fact slaves (processing data) that fetch input data, merge and store results.
2. After fetching an inputoutput lock, each thread fetches input data, if available.
3. The thread processes the input.
4. After fetching the lock again, each thread stores its result in the output buffer and checks whether a sufficient number of results is available. If this is the case then the thread merges results. Then it checks whether new input should be generated. If this is the case then the thread generates new data by invoking function `generate_new_input(...)`.

This way master and slave functions are merged. The main part of the code implementing the improved approach is shown in Listing 5.3 while full code is available in file `master-slave-OpenMP-1.c`.

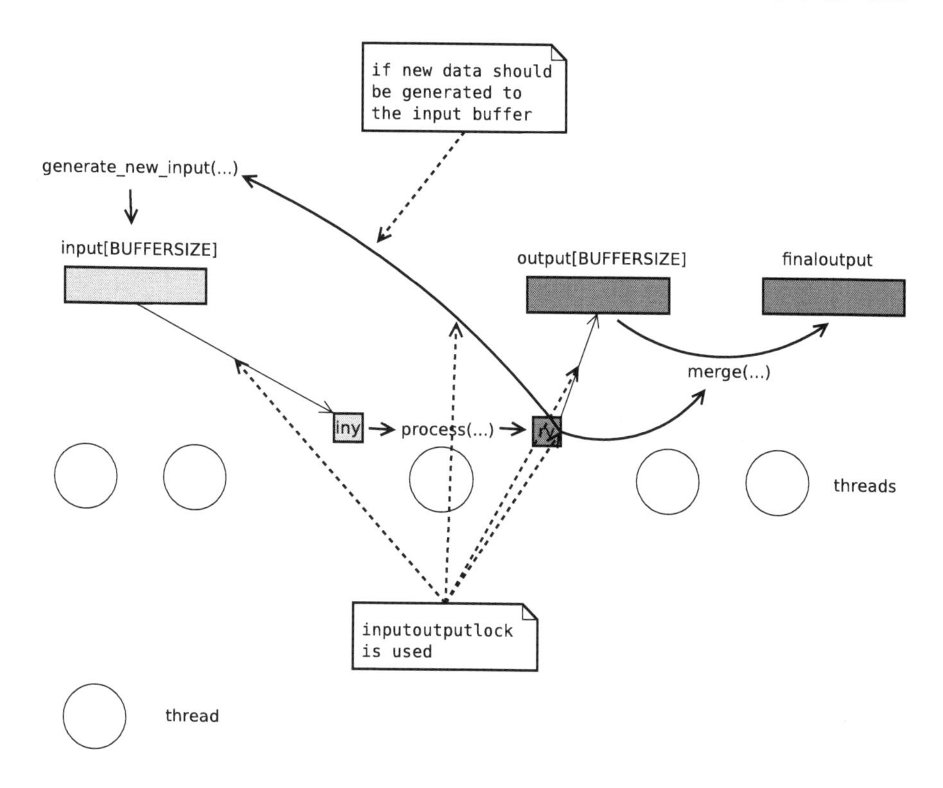

FIGURE 5.4 Master-slave scheme with OpenMP – version with all threads executing same code

Listing 5.3 Master-slave in OpenMP with each thread fetching input data and storing output – main part

```
  omp_init_lock(&inputoutputlock);
  // firstly generate BUFFERSIZE data chunks of input data
  lastgeneratedcount=generate_new_input(input);
  init_final_output(&finaloutput);

#pragma omp parallel private(i) shared(input,output,finaloutput,
    currentinputindex,currentoutputcount,lastgeneratedcount,
    processedcount) num_threads(threadnum)
  {
    // each thread acts as a slave
    int processdata;
    t_output result;
    long myinputindex;
    int finish;
```

```
    do {
      processdata=0;
      finish=0;
      omp_set_lock(&inputoutputlock);
      if (processedcount<CHUNKCOUNT) {
        myinputindex=currentinputindex;
        if (currentinputindex<lastgeneratedcount) {
          currentinputindex++;
          processdata=1;
        }
      } else finish=1;
      omp_unset_lock(&inputoutputlock);

      if (processdata) {
        // now process the input data chunk
        result=process(&(input[myinputindex]));

        // store the result in the output buffer
        omp_set_lock(&inputoutputlock);
        if (currentoutputcount<BUFFERSIZE) {
          output[currentoutputcount]=result;
          currentoutputcount++;
        }
        if (currentoutputcount==lastgeneratedcount) { // process
    all available results
          for(i=0;i<lastgeneratedcount;i++)
            merge(&finaloutput,&(output[i]));
          processedcount+=lastgeneratedcount;
          currentoutputcount=0;

          if (processedcount<CHUNKCOUNT) {
            lastgeneratedcount=generate_new_input(input);
            currentinputindex=0;
          }
        }
        omp_unset_lock(&inputoutputlock);
      }
    } while (!finish);
  }
  print_final_output(&finaloutput);
  omp_destroy_lock(&inputoutputlock);
```

Finally, we analyze a third version of an OpenMP code available in file `master-slave-OpenMP-2.c`, with the idea shown in Figure 5.5 and its main part presented in Listing 5.4. It uses the `#pragma omp task` directive, described in Section 4.2.9. Specifically, within a `#pragma omp parallel` block one thread:

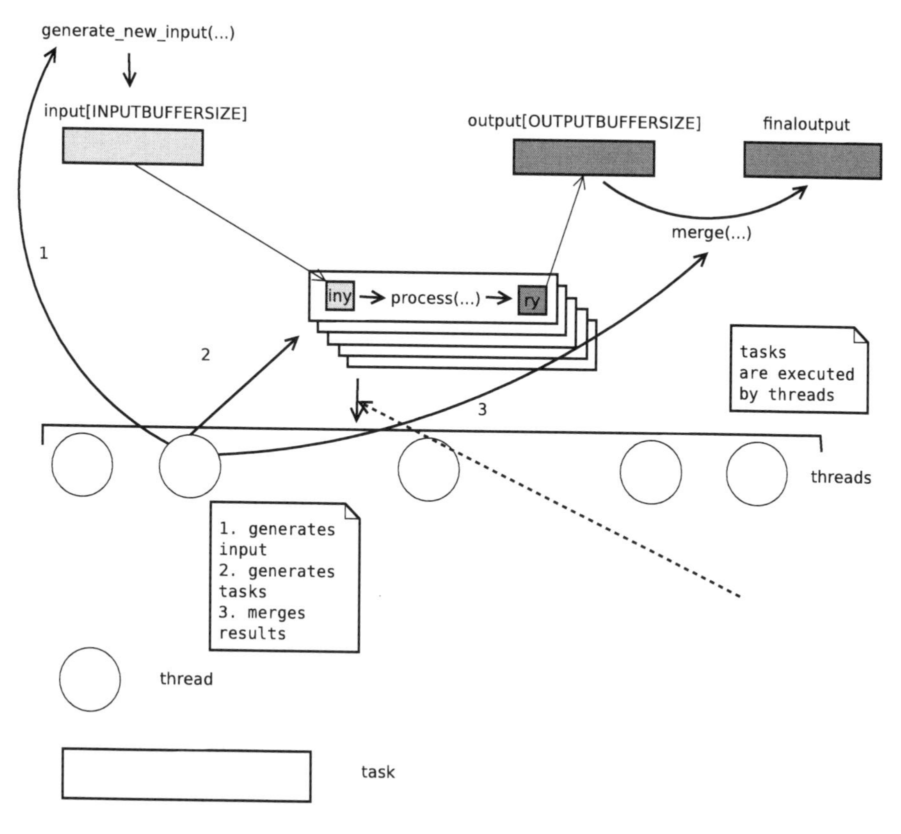

FIGURE 5.5 Master-slave scheme with OpenMP – version with tasks

1. Generates input data by invoking function `generate_new_input(...)`.
2. Generates a number of tasks each of which will deal with processing of a specific input indicated by `myinputindex`, these will be processed by threads.
3. Waits for results (`#pragma omp taskwait`) and merges results into a final output.
4. Repeats the above steps until all chunks have been processed.

Listing 5.4 Master-slave in OpenMP using the `#pragma omp task` directive – main part

```
init_final_output(&finaloutput);
```

```
#pragma omp parallel private(i,myinputindex) shared(input,output
    ,finaloutput,lastgeneratedcount) num_threads(threadnum)
  {
#pragma omp single
    {
      long processedcount=0;
      do {
        lastgeneratedcount=generate_new_input(input);
        // now create tasks that will deal with data packets
        for(myinputindex=0;myinputindex<lastgeneratedcount;
    myinputindex++)
          {
#pragma omp task firstprivate(myinputindex) shared(input,output)
            {
              // now each task is processed independently and
    can store its result into an appropriate buffer
              output[myinputindex]=process(&(input[myinputindex
    ]));
            }
          }
        // wait for tasks
#pragma omp taskwait
        // now merge results
        for(i=0;i<lastgeneratedcount;i++)
          merge(&finaloutput,&(output[i]));
        processedcount+=lastgeneratedcount;
      } while (processedcount<CHUNKCOUNT);
    }
  }
  print_final_output(&finaloutput);
}
```

The codes can be compiled as follows:

```
gcc -fopenmp <flags> master-slave-OpenMP-<0,1,2>.c -lm
```

and executed by running the executable – `./a.out` in this case.

Table 5.1 presents execution times for the three versions (best out of three runs for each configuration are reported) on a workstation with 2 x Intel Xeon E5-2620v4 and 128 GB RAM. It can be seen that each successive version resulted in better performance compared to a previous one.

TABLE 5.1 Execution times [s] for various versions of the master-slave OpenMP code

Code version	Execution time [s]
with a designated master thread	33.936
each thread fetches input data and stores output	33.521
using the #pragma omp task directive	32.630

5.1.3 MPI+OpenMP

The code presented in Section 5.1.1 can be extended for parallelization with OpenMP within each process. Since multicore processors are widely available, a combination of MPI for communication between nodes and OpenMP for parallelization of slave computations is a good approach. Such an approach is presented in Figure 5.6.

Proposed code, available in file master-slave-MPI+OpenMP.c differs from the MPI only approach from Section 5.1.1 in the following:

1. MPI initialization is performed using MPI_Init_thread(...) instead of MPI_Init(...) for multithreaded support of level MPI_THREAD_FUNNELED:

```
MPI_Init_thread(&argc, &argv,MPI_THREAD_FUNNELED,&
    threadsupport);
if (threadsupport<MPI_THREAD_FUNNELED) {
  printf("\nThe implementation does not support
MPI_THREAD_FUNNELED, it supports level %d\n",threadsupport)
    ;
  MPI_Finalize();
  return -1;
}
```

For MPI_THREAD_FUNNELED the master thread is allowed to call MPI functions. For MPI+OpenMP implementations this means calling from outside #pragma OpenMP regions, within the #pragma omp master clause or within a code in a process for which MPI_Is_thread_main(int *flag) returned a true flag [15].

2. Parallelization within each node can be done within function slavecpu(...) using the #pragma omp parallel directive. For good performance, this approach assumes that computations are balanced well among threads at this stage. Listing 5.5 presents function slavecpu(...) within file master-slave-MPI+OpenMP.c.

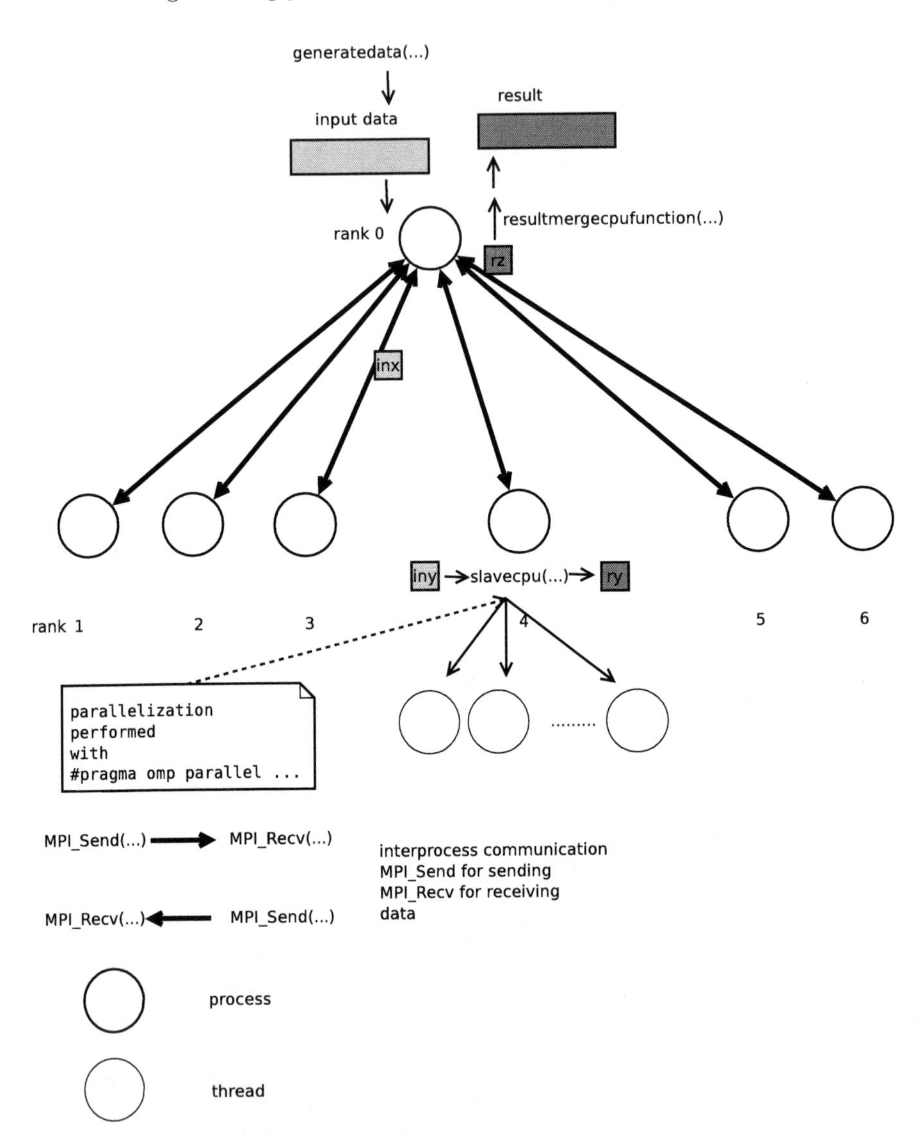

FIGURE 5.6 Master-slave scheme with MPI and OpenMP

Listing 5.5 Basic master-slave application using MPI and OpenMP – function `slavecpu(...)`

```
void slavecpu(double *result,double start,double end,double step
    ,int threadcount) {
  double sresult=0;
  long counter;
```

```
  long countermax;
  double arg;
  int threadid;

#pragma omp parallel private(threadid,countermax,arg,counter)
    shared(start,end,step,threadcount) reduction(+:sresult)
    num_threads(threadcount)
{
  threadid=omp_get_thread_num();
  countermax=((end-start)/step)/threadcount;

  arg=start+step*threadid;
  for(counter=0;counter<countermax;counter++) {
    sresult+=step*(fcpu(arg)+fcpu(arg+step))/2;
    arg+=threadcount*step;
  }
}
  *result=sresult;
}
```

The code can be compiled as follows:

```
mpicc -fopenmp <flags> master-slave-MPI+OpenMP.c
```

and subsequently run in the following way, as an example:

```
mpirun -np 3 ./a.out 15
```

In this particular case there is 1 master process and 2 slave processes each of which uses OpenMP for parallel processing with 15 threads.

Table 5.2 presents results obtained from a testbed run (here MPI communication takes places within one node) on a workstation with 2 x Intel Xeon E5-2620v4 and 128 GB RAM for various configurations. Results show scalability of computations. Presented results for each configuration are best times out of three runs.

5.1.4 MPI+Pthreads

The traditional master-slave scheme can also be implemented with MPI and Pthreads for hybrid environments i.e. clusters with multiple cores within each node. MPI can be preferably used for communication between nodes while Pthreads can be used to manage input data, computations and results across threads that run on cores within a node. In contrast to the MPI+OpenMP version, the MPI+Pthreads solution shown in Figure 5.7 is much more flexible but also much more complex. Many threads with various roles are used within processes. Key code of the master process is shown in Listing 5.6. The implementation uses the following threads within each slave process:

TABLE 5.2 Execution time [s] for various configurations of the MPI+OpenMP code

number of processes – master+slave(s)	number of computing threads per slave process	Execution time [s]
3	1	64.313
3	2	33.331
3	4	18.542
3	16	11.450

- master (main) thread – used to receive data packets which are stored and then fetched by computing threads – key code shown in Listing 5.7,
- computing threads – used to process incoming data packets – key code shown in Listing 5.8,
- sending thread – one thread used to send processed data packets – key code shown in Listing 5.9; the thread sends a temporary result back to the master process which merges results from various processes.

The implementation assumes the following data holders:

1. Input queue within each process – holds input data packets that are received from the master using MPI. Data packets from the input queue can be then fetched by computing threads within the process.
2. Output queue within each process – used to store results of already processed data packets. Each computing thread stores result of its data packet after which it fetches a new data packet to process.

Consequently, the implementation allows the following:

1. Efficient processing of data packets that take various amounts of time to process as threads do not need to perform computations synchronously and are not synchronized as a group.
2. Decoupling sending results back from processing next data packets. If the main thread has already uploaded enough data packets, then a compute a thread, upon finishing processing of a previous data packet and uploading its result to the output queue, can start processing of a new data packet immediately, while the sending thread takes care of sending the result back.

In order to implement such a solution, an MPI implementation must support the `MPI_THREAD_MULTIPLE` mode since in the slave process the main thread would receive new data packets while results are sent back to the master process by the sending thread.

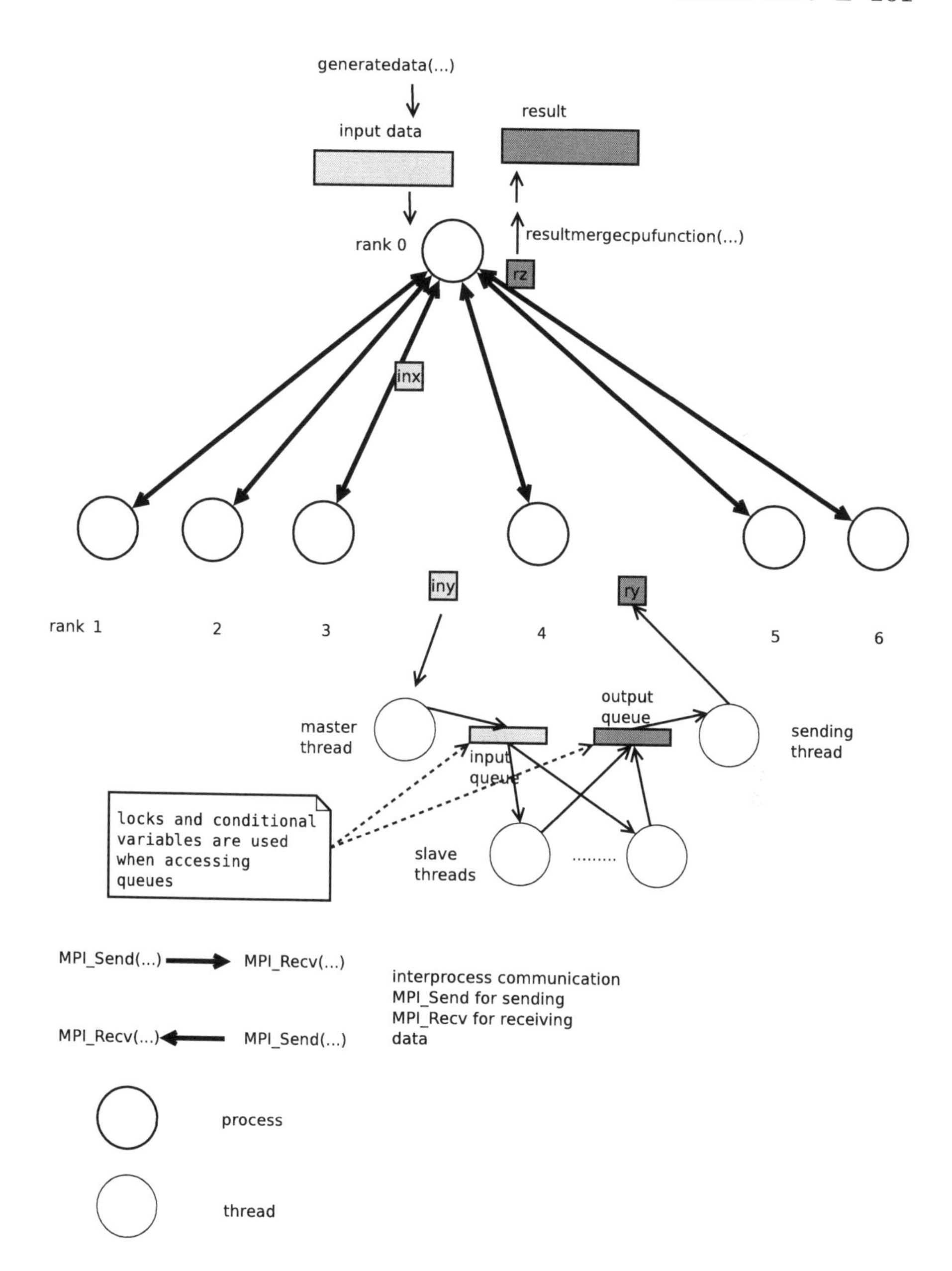

FIGURE 5.7 Master-slave scheme with MPI and Pthreads

Furthermore, the implementation requires synchronization among threads in each slave process. Specifically, the producer-consumer pattern is implemented in two places – for:

1. `inputqueue` – the main thread receiving new data packets acts as the producer while computing threads act as consumers.
2. `outputqueue` – computing threads act as producers and the sending thread acts as a consumer.

For these implementations, mutexes and conditional variables are used. The latter allow a thread to wait on a conditional variable for an event to occur (such as a queue to have place to put new elements or to have elements to fetch).

Furthermore, it should be noted that this implementation, while very flexible and with a potential for load balancing, will likely result in more overhead for well balanced computations where various data packets could be computed at the same time by various computing threads. In such a case, there would be no need for queues and an implementation such as the one presented in Section 5.1.3 would be very efficient.

Listing 5.6 Master-slave application using MPI and Pthreads and input and output queues – key master code

```
long packetcounter=0;
// distribute initial data packets to slave processes to feed
    threads
for(j=0;j<cputhreadcount;j++)
  for(i=1;i<processcount;i++,packetcounter++)
    MPI_Send(&(data[2*packetcounter]),2,MPI_DOUBLE,i,DATA_PACKET
    ,MPI_COMM_WORLD);

// now wait for results and respond to incoming result messages
// do it individually, in general data packets may take more or
    less time
// to process (e.g. due to various CPUs on various nodes)
do {
  MPI_Recv(&temporaryresult,1,MPI_DOUBLE,MPI_ANY_SOURCE,RESULT,
    MPI_COMM_WORLD,&mpistatus);
  // merge result
  resultmergecpufunction(&args);              // result in variable
    result

  // check who sent the result and send another data packet to
    that process
  MPI_Send(&(data[2*packetcounter]),2,MPI_DOUBLE,mpistatus.
    MPI_SOURCE,DATA_PACKET,MPI_COMM_WORLD);
```

```
  packetcounter++;

} while (packetcounter<packetcount); // do it as long as there
    are data packets to send

// now wait for pending results
for(j=0;j<cputhreadcount;j++)
  for(i=1;i<processcount;i++) {
    MPI_Recv(&temporaryresult,1,MPI_DOUBLE,MPI_ANY_SOURCE,RESULT
    ,MPI_COMM_WORLD,&mpistatus);
    resultmergecpufunction(&args);        // result in variable
    result
  }
// and send a termination message
for(i=1;i<processcount;i++)
  MPI_Send(NULL,0,MPI_DOUBLE,i,FINISH_COMPUTATIONS,
    MPI_COMM_WORLD);
// display the result
printf("\nResult is %f\n",result);
fflush(stdout);
```

Listing 5.7 Master-slave application using MPI and Pthreads and input and output queues – key main thread code of a slave

```
(...)
// the main thread will manage MPI communication, insert data
    packets into the queue and merge results
double data[2];
do {
  MPI_Probe(0,MPI_ANY_TAG,MPI_COMM_WORLD,&mpistatus);
  if (mpistatus.MPI_TAG==DATA_PACKET) {
    MPI_Recv(data,2,MPI_DOUBLE,0,DATA_PACKET,MPI_COMM_WORLD,&
    mpistatus);

    // add the data packet to the queue
    pthread_mutex_lock(&inputqueuemutex);
    while (inputqueuesize+1>QUEUE_SIZE)
      pthread_cond_wait(&inputqueuecondvariableproducer,&
    inputqueuemutex);

    long index=2*inputqueuesize;     // input queue size refers
    to the number of ranges
    inputqueue[index++]=data[0];
    inputqueue[index++]=data[1];
    inputqueuesize++;
```

```
    inputdatapacketrequestedcounter++;

    pthread_cond_signal(&inputqueuecondvariableconsumer);
    pthread_mutex_unlock(&inputqueuemutex);

    pthread_mutex_lock(&outputqueuemutex);
    outputdatapacketrequestedcounter++;
    pthread_mutex_unlock(&outputqueuemutex);
  }
} while (mpistatus.MPI_TAG!=FINISH_COMPUTATIONS);

// ask threads (that may be waiting on waits) to finish
(...)

// wait for completion of threads
pthread_join(mergingthread,&threadstatus);
for (i=0;i<cputhreadcount;i++)
  pthread_join(thread[i],&threadstatus);
(...)
```

Listing 5.8 Master-slave application using MPI and Pthreads and input and output queues – key compute slave code

```
double sresult=0;
char threadfinish=0;
while (!threadfinish) {

    // try to get a data packet from the input queue
  pthread_mutex_lock(&inputqueuemutex);
  while ((!inputthreadfinish) && (inputqueuesize==0)) {
    pthread_cond_wait(&inputqueuecondvariableconsumer,&
    inputqueuemutex);
  }

  if ((inputthreadfinish) && (inputdatapacketrequestedcounter==
    inputdatapacketprocessedcounter)) threadfinish=1; //
    threadfinish should
  // be "true" only if all requested data packets have been
    processed

  fetcheddatapacket=0;
  if ((inputqueuesize>0))  {
    inputqueuesize--;
    start=inputqueue[2*inputqueuesize];
    end=inputqueue[2*inputqueuesize+1];
```

```
    inputdatapacketprocessedcounter++;
    fetcheddatapacket=1;
  }

  pthread_cond_signal(&inputqueuecondvariableproducer);
  pthread_mutex_unlock(&inputqueuemutex);

  if (fetcheddatapacket) {
    // set the step
    step=(end-start)/(40*DATA_PACKET_PARTITION_COUNT);
    sresult=0;
    // process the data packet
    countermax=((end-start)/step);
    arg=start;
    for(counter=0;counter<countermax;counter++) {
        sresult+=step*(fcpu(arg)+fcpu(arg+step))/2;
        arg+=step;
    }
  }

  if (fetcheddatapacket) {
    // insert result into the final queue
    pthread_mutex_lock(&outputqueuemutex);
    while (outputqueuesize==QUEUE_SIZE)
      pthread_cond_wait(&outputqueuecondvariableproducer,&
    outputqueuemutex);

    outputqueue[outputqueuesize]=sresult;
    outputqueuesize++;

    pthread_cond_signal(&outputqueuecondvariableconsumer);
    pthread_mutex_unlock(&outputqueuemutex);
  }
}
```

Listing 5.9 Master-slave application using MPI and Pthreads and input and output queues – key sending thread code of a slave

```
char threadfinish=0;

while (!threadfinish) {

  // get results from the output queue
  pthread_mutex_lock(&outputqueuemutex);
  while ((!outputthreadfinish) && (outputqueuesize==0)) {
```

```
    pthread_cond_wait(&outputqueuecondvariableconsumer,&
    outputqueuemutex);
  }
  if ((outputthreadfinish) && (outputdatapacketrequestedcounter
    ==outputdatapacketprocessedcounter)) threadfinish=1;
  // terminate only after all requested packets have been
    processed

  fetcheddatapacket=0;
  if (outputqueuesize>0) {
    // copy output
    temporaryresult=outputqueue[--outputqueuesize];
    outputdatapacketprocessedcounter++;
    fetcheddatapacket=1;
  }
  pthread_cond_signal(&outputqueuecondvariableproducer);
  pthread_mutex_unlock(&outputqueuemutex);

  if (fetcheddatapacket) {
    MPI_Send(&temporaryresult,1,MPI_DOUBLE,0,RESULT,
    MPI_COMM_WORLD);
  }
}
```

The code can be compiled as follows:

```
mpicc <flags> master-slave-MPI+Pthreads.c
```

and can be run in the following way, as an example

```
mpirun -np 3 ./a.out 8
```

which runs 1 master, 2 slave processes each of which starts 8 threads.

Results obtained on on a workstation with 2 x Intel Xeon E5-2620v4 and 128 GB RAM are shown in Table 5.3.

TABLE 5.3 Execution time [s] for various configurations of the MPI+Pthreads code

number of processes – master+slave(s)	number of computing threads per slave process	Execution time [s]
3	1	65.681
3	2	35.289
3	4	19.224
3	16	11.594

It should be noted that, compared to the previously shown MPI implementation here the configuration with 1 computing thread per slave process might actually involve an additional overhead for thread creation.

5.1.5 CUDA

Figure 5.8 presents a basic master-slave scheme that can use many GPUs installed within a node. This version uses one host thread and the CUDA API for management of computations among GPUs. Specifically, input data is partitioned into several data chunks which are then distributed among GPUs. This particular example uses numerical integration of function `f(...)` over range [`start,end`]. In this application basic data structures and operations are as follows:

- a data packet corresponds to a subrange of the initial range,
- for each data packet a step is generated that determines into how many further subranges it is partitioned for subsequent integration using the trapezoid method.

It should be noted that the merging phase is actually a two-step process:

1. merging results computed in various thread blocks,
2. merging results of various data packets.

Depending on the application, these steps might be very quick compared to the execution time of computing actual data packets. It is possible to overlap these merging activities with computation of successive data packets, but depending on relative execution times, this might not bring considerable or visible benefits in execution time. In such a case, a simple implementation balancing load among computing devices might be sufficient. In the implementation for numerical integration included in file `master-slave-n-GPUs-one-thread-sequential-cpu-merging.cu` merging results for various thread blocks is done sequentially on the host. In this application this step is very fast. If, on the other hand, it takes much time, running it in parallel, potentially overlapping with spawning new kernels on the GPU, is possible. Computations contained in the computational kernel in this code are shown in Listing 5.10.

Listing 5.10 Basic master-slave application using multiple GPUs with CUDA and one host thread – kernel code

```
__global__
void slavegpu(double *result,double start,double end,double step
    ) {
```

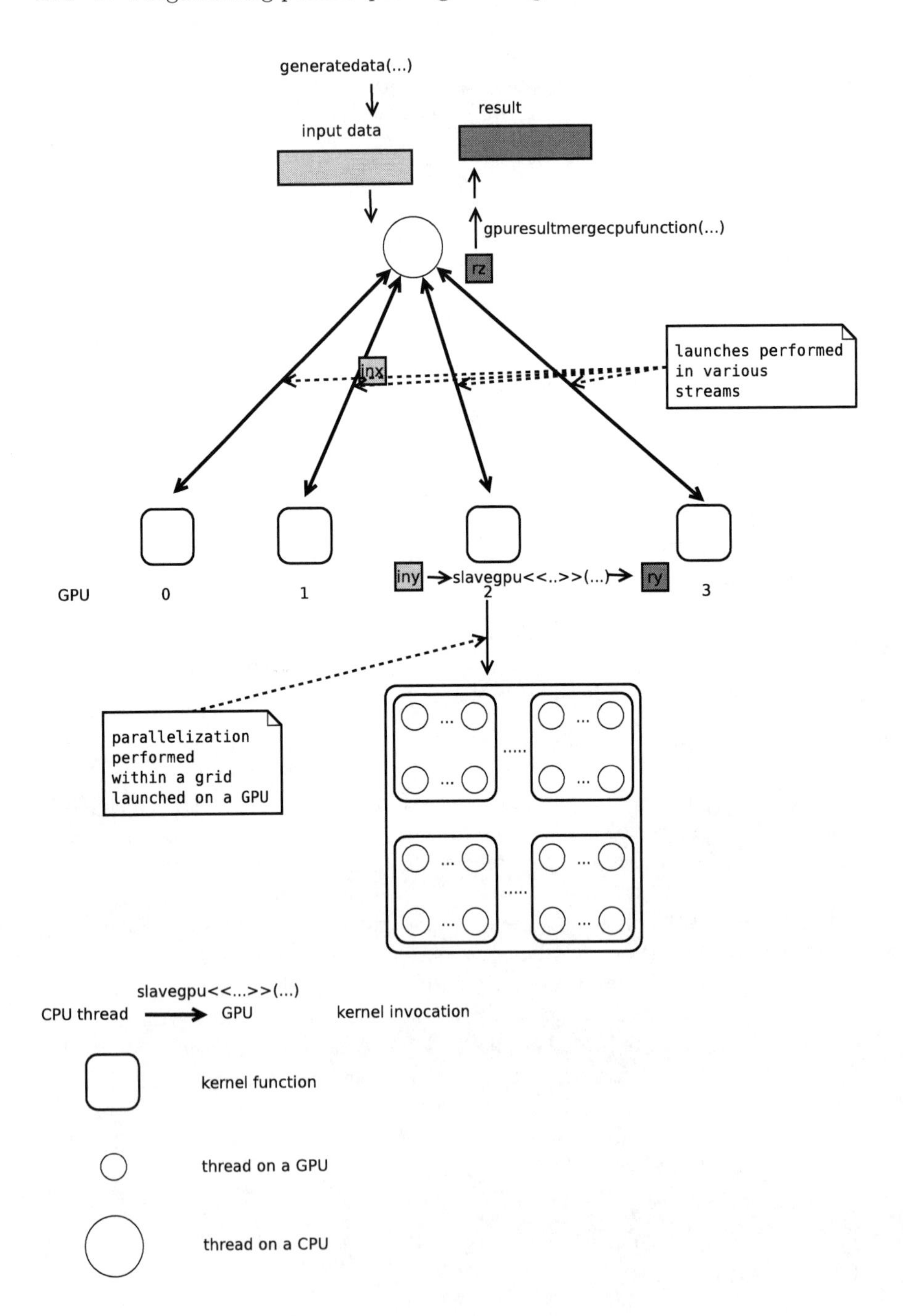

FIGURE 5.8 Master-slave scheme with CUDA

```
  extern __shared__ double sresults[]; // use dynamically
    allocated shared memory
  long threadcount;
  long counter;
  long countermax;
  double arg;

  threadcount=gridDim.x*blockDim.x;
  countermax=((end-start)/step)/threadcount;
  arg=start+step*(blockIdx.x*blockDim.x+threadIdx.x);
  sresults[threadIdx.x]=0;
  for(counter=0;counter<countermax;counter++) {
    sresults[threadIdx.x]+=step*(f(arg)+f(arg+step))/2;
    arg+=threadcount*step;
  }

  __syncthreads();

  for(counter=512;counter>0;counter/=2) {
    if (threadIdx.x<counter)
      sresults[threadIdx.x]+=sresults[threadIdx.x+counter];
    __syncthreads();
  }

  // store the block maximum in a global memory
  if (threadIdx.x==0)
    result[blockIdx.x]=sresults[0];
}
```

It should be noted that this parallel code would typically invoke kernels on each GPU several times as shown in the host thread code in Listing 5.11. It has the potential of load balancing even in heterogeneous environments in which there are several GPUs with various performance. Effective load balancing is possible if:

1. the number of data packets is large enough to hide idle times of faster GPUs by giving more work while slower GPUs still process previous data chunks,

2. execution time of a data chunk is considerable compared to kernel launch, data copy and synchronization overheads.

Listing 5.11 Basic master-slave application using multiple GPUs with CUDA and one host thread – host thread key code

```
for(i=0;i<requesteddevicecount;i++) {
```

```
  gpuactive[i]=0;
  hresults[i]=(double *)malloc(size);
  if (!(hresults[i])) errorexit("Error allocating memory on the
    host");

  cudaSetDevice(i);

  if (cudaSuccess!=cudaMalloc((void **)&(dresults[i]),size))
    errorexit("Error allocating memory on the GPU");

  if (cudaSuccess!=cudaStreamCreate(&(stream[i])))
    errorexit("Error creating stream");

  // check if there are more and get a new data chunk
  if (packetcounter<packetcount) {
    start[i]=data[packetcounter++];
    end[i]=data[packetcounter];
    step=(end[i]-start[i])/(40*blocksingrid*threadsinblock);
    // start computations on the GPU using particular streams
    // use dynamically allocated shared memory
    slavegpu<<<blocksingrid,threadsinblock,threadsinblock*sizeof
    (double),stream[i]>>>(dresults[i],start[i],end[i],step);
    if (cudaSuccess!=cudaGetLastError())
      errorexit("Error during kernel launch in stream");
    gpuactive[i]=1;
  } else
    errorexit("Too few data packets");
}

// now wait for some result
do {
  finish=1;
  for(i=0;i<requesteddevicecount;i++) {
    cudaSetDevice(i);
    if ((gpuactive[i]) && (cudaSuccess==cudaStreamQuery(stream[i
    ]))) { // computations completed
      // copy results from a GPU to the host
      if (cudaSuccess!=cudaMemcpyAsync(hresults[i],dresults[i],
    size,cudaMemcpyDeviceToHost,stream[i]))
        errorexit("Error copying results");
      cudaStreamSynchronize(stream[i]);

      startvalue[i].elemcount=blocksingrid;
      startvalue[i].elements=hresults[i];
      startvalue[i].result=&(functionresults[i]);
```

```
      gpuresultmergecpufunction(&(startvalue[i]));
      finalresult+=*(startvalue[i].result);

      // spawn new work on the GPU if there is more work to do
      if (packetcounter<packetcount) {
        start[i]=data[packetcounter++];
        end[i]=data[packetcounter];
        step=(end[i]-start[i])/(40*blocksingrid*threadsinblock);
        slavegpu<<<blocksingrid,threadsinblock,threadsinblock*
    sizeof(double),stream[i]>>>(dresults[i],start[i],end[i],step
    );
        if (cudaSuccess!=cudaGetLastError())
          errorexit("Error during kernel launch in stream1");
        gpuactive[i]=1;
      } else gpuactive[i]=0;
    }
    if (gpuactive[i]==1) finish=0; // if at least one GPU is
    processing then do not finish
  }
} while (!finish);

printf("\nThe final result is %f\n",finalresult);
// release resources
(...)
```

Nevertheless, there is overhead for kernel launch. It is interesting to assess how performance differs when comparing the following two configurations:

1. configuration with `DATA_PACKET_COUNT=1` used in data generation code:

```
__host__
void generatedata(double **data,long *packetcount) {
  double start=1,end=1000000;
  double counter=start;
  double step=(end-start)/(double)DATA_PACKET_COUNT;
  long i;

  *data=(double *)malloc(sizeof(double)*(1+
    DATA_PACKET_COUNT));
  if (!(*data))
     errorexit("Not enough memory on host when generating
    data");

  for(i=0;i<=DATA_PACKET_COUNT;i++,counter+=step)
    (*data)[i]=counter;
```

```
    *packetcount=DATA_PACKET_COUNT;
  }
```

   and a certain size of `step` for partitioning within each data packet shown in Listing 5.10,

2. configuration with an increased value of `DATA_PACKET_COUNT` e.g. increased n times while the value of `step` is decreased n times.

Both versions result in the same width of a subrange used for the trapezoid method within the GPU kernel but differ in the number of data packets. It is expected that a larger number of data packets would generate an overhead. Results obtained on on a workstation with 2 x Intel Xeon E5-2620v4, 2 x NVIDIA GTX 1070 and 128 GB RAM are shown in Table 5.4. For each configuration the best time out of three runs is presented. Two GPUs were used.
The code can be compiled as follows:

```
nvcc master-slave-n-GPUs-one-thread-sequential-cpu-merging.cu
```

and then run in the following way for 2 GPUs:

```
./a.out 2
```

TABLE 5.4 Execution time [s] for various configurations of the CUDA code

DATA_PACKET_COUNT	coefficient used in the denominator for computing step	Execution time [s]
2	2000	3.992
10	400	4.020
100	40	4.118
1000	4	5.210

It is possible to implement a queue based scheme in which the host thread in charge of GPU launch management transfers results from GPUs to a queue from which threads in a pool fetch data and process. This can be implemented in Pthreads using mutexes and conditional variables, similarly to the producer-consumer scheme [128, Chapter 4].

One of disadvantages of this code is the way the host thread manages work. Specifically, as status of GPUs is checked, the host thread in fact implements busy waiting. This can be improved by launching several threads for management of distinct GPUs as shown in Section 5.1.6. Such an approach potentially allows utilizing CPU cores for processing data packets.

5.1.6 OpenMP+CUDA

Another implementation uses many host threads to manage computations on GPU(s) and host CPU(s). Specifically, the idea of this solution, presented in Figure 5.9 is as follows:

1. Input data is partitioned into data packets as in the previous code.
2. For each GPU there is a separate thread managing computations on a corresponding GPU. For host CPU(s) there is one thread managing computations. Each of these threads would fetch an id of the next data packet in an OpenMP critical section and start computations in parallel on a device (using CUDA for GPU(s) and OpenMP for CPU(s)) and wait for results in a blocking way. The latter can be done with `cudaStreamSynchronize(...)` as the host threads work independently and synchronize only on the critical section. It is important that each thread within an OpenMP parallel region sets the active CUDA device [109]. Actual key code for the managing threads is shown in Listing 5.12.
3. Computational code is performed within a kernel on a GPU – analogous to the code shown in Listing 5.10 before and within function `slavecpu(...)` using OpenMP for parallelization and shown in Listing 5.13.

Listing 5.12 Basic master-slave application using multiple GPUs with CUDA and CPU with OpenMP and multiple host threads – key host thread code

```
  for(i=0;i<requesteddevicecount;i++) {
    hresults[i]=(double *)malloc(size);
    if (!(hresults[i])) errorexit("Error allocating memory on
    the host");

    cudaSetDevice(i);

    if (cudaSuccess!=cudaMalloc((void **)&(dresults[i]),size))
        errorexit("Error allocating memory on the GPU");

    if (cudaSuccess!=cudaStreamCreate(&(stream[i])))
       errorexit("Error creating stream");
  }

#pragma omp parallel private(finish,mypacketid,step) shared(
    blocksingrid,threadsinblock,packetcounter,packetcount,start,
    end,dresults,stream,hresults,size) reduction(+:finalresult)
    num_threads(requesteddevicecount+((cputhreadcount>0)?1:0))
```

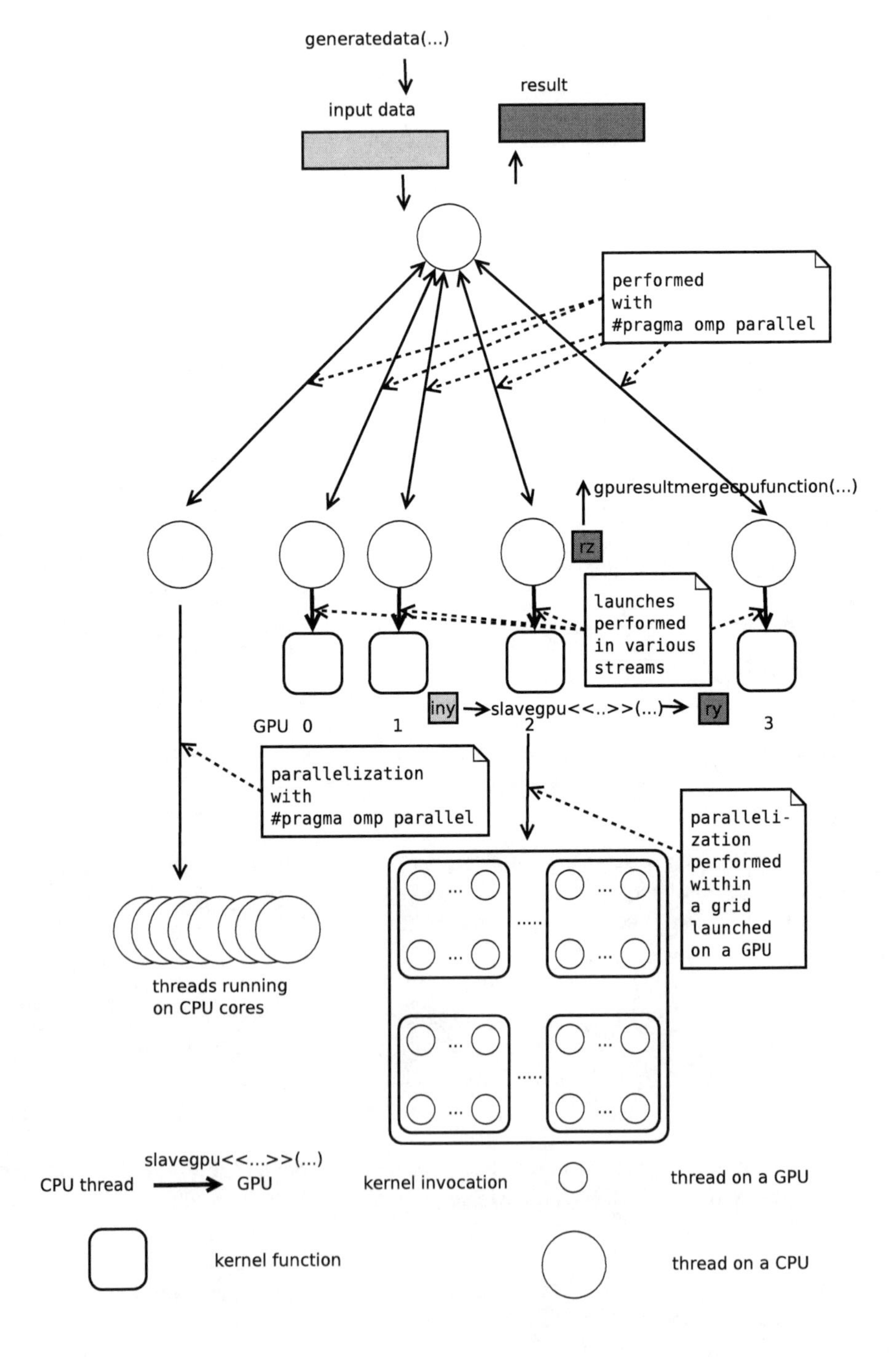

FIGURE 5.9 Master-slave scheme with OpenMP and CUDA

```
{
  finalresult=0;
  int i=omp_get_thread_num();

  if (i<requesteddevicecount) { // threads managing execution
  on GPU(s)
    cudaSetDevice(i);
    do {
      finish=0;
      #pragma omp critical
      {
        // check if there are more and get a new data chunk
        if (packetcounter<packetcount)
          mypacketid=packetcounter++;
        else
          finish=1;
      }
      if (finish==0) {
        start[i]=data[mypacketid];
        end[i]=data[mypacketid+1];
        step=(end[i]-start[i])/(40*blocksingrid*threadsinblock
 );

        // start computations on the GPU using particular
 streams
        // use dynamically allocated shared memory
        slavegpu<<<blocksingrid,threadsinblock,threadsinblock*
 sizeof(double),stream[i]>>>(dresults[i],start[i],end[i],step
 );
        if (cudaSuccess!=cudaGetLastError())
          errorexit("Error during kernel launch in stream");

        // copy results from a GPU to the host
        if (cudaSuccess!=cudaMemcpyAsync(hresults[i],dresults[
 i],size,cudaMemcpyDeviceToHost,stream[i]))
          errorexit("Error copying results");
        // wait for completion of processing on the GPU
        cudaStreamSynchronize(stream[i]);

        // and merge results
        startvalue[i].elemcount=blocksingrid;
        startvalue[i].elements=hresults[i];
        startvalue[i].result=&(functionresults[i]);
        gpuresultmergecpufunction(&(startvalue[i]));
```

```
          finalresult+=*(startvalue[i].result);
        }
      } while (!finish);
    } else { // parallel execution on CPU(s)
      do {
        finish=0;
        #pragma omp critical
        {
          // check if there are more and get a new data chunk
          if (packetcounter<packetcount)
            mypacketid=packetcounter++;
          else
            finish=1;
        }
        if (finish==0) {
          start[i]=data[mypacketid];
          end[i]=data[mypacketid+1];
          step=(end[i]-start[i])/(40*blocksingrid*threadsinblock
    );
          // start computations on the CPU(s) -- the result will
     be one double in this case
          slavecpu(&(functionresults[i]),start[i],end[i],step,
    cputhreadcount);
          // merge results
          finalresult+=functionresults[i];
        }
      } while (!finish);
    }
  }
  printf("\nThe final result is %f\n",finalresult);
  // release resources
  (...)
```

Listing 5.13 Basic master-slave application using multiple GPUs with CUDA and CPU with OpenMP and multiple host threads – `slavecpu(...)` function

```
void slavecpu(double *result,double start,double end,double step
    ,int threadcount) {
  double sresult=0;
  long counter;
  long countermax;
  double arg;
  int threadid;
```

```
#pragma omp parallel private(threadid,countermax,arg,counter)
    shared(start,end,step,threadcount) reduction(+:sresult)
    num_threads(threadcount)
{
  threadid=omp_get_thread_num();
  countermax=((end-start)/step)/threadcount;

  arg=start+step*threadid;
  for(counter=0;counter<countermax;counter++) {
    sresult+=step*(fcpu(arg)+fcpu(arg+step))/2;
    arg+=threadcount*step;
  }
}
  *result=sresult;
}
```

The program invocation has the following syntax:

```
Syntax: program <the number of GPUs to use> \
<the number of CPU threads for computing>
```

and can be compiled as follows in the Linux environment:

```
nvcc -Xcompiler -fopenmp -Xlinker -lgomp \
master-slave-n-GPUs-multiple-threads-OpenMP.cu \
-o master-slave-n-GPUs-multiple-threads-OpenMP
```

The `-Xcompiler` option allows to pass options directly to the compiler encapsulated by `nvcc` while `-Xlinker` allows to pass options to the host linker.

In terms of performance and speed-ups, the following can be noted with respect to the implementation:

1. It is possible to specify 1+GPUs/0+ CPU computing threads or 0+ GPUs/1+ CPU computing threads to use. This refers to CPU threads that perform slave jobs.

2. Typical notes regarding speed-up would also apply to this case i.e.:

   - larger input data size would result in higher speed-up,
   - decreasing the number of data packages while increasing the number of subranges in each of them, as suggested in Section 5.1.5 would result in higher speed-up.

3. The implementation scales in a hybrid environment.

Following tests show execution times for various configurations in terms of the number of computing threads on CPU(s) and the number of GPUs involved and were performed on a workstation with 2 x Intel Xeon E5-2620v4, 2 x NVIDIA GTX 1070 and 128 GB RAM. `DATA_PACKET_COUNT=100` and `40*` were used in the denominator for computing `step`.

TABLE 5.5 Execution time [s] for various configurations of the hybrid OpenMP+CUDA code

number of CPU threads	number of GPUs involved	Execution time [s]
1	0	379.277
2	0	213.16
4	0	129.871
8	0	75.384
16	0	53.779
32	0	35.458
30	1	6.273
28	2	3.954
30	2	3.877
32	2	3.923

Additionally, tests were performed for various values of `DATA_PACKET_COUNT` and the coefficient used in the denominator for computing `step`. Results are presented in Table 5.6. It can be seen that too few packets do not allow the use of available computing cores and too large a number of data packets results in additional overhead. Tests were run on the same platform as described above.

TABLE 5.6 Execution time [s] for various configurations of the hybrid OpenMP+CUDA code, 30 CPU threads, 2 GPUs used

`DATA_PACKET_COUNT`	coefficient used in the denominator for computing `step`	Execution time [s]
10	400	4.805
100	40	3.877
1000	4	4.743

5.2 GEOMETRIC SPMD

5.2.1 MPI

This section discusses an MPI+C code for parallel implementation of the geometric SPMD scheme, included in file `SPMD-MPI-1.c`. A 3D domain is assumed which is partitioned into subdomains in function `optimizepartitioning(...)` according to the algorithm presented in Section

3.3. Impact of various partitioning schemes, including stripe based partitioning and the one implemented in the code discussed in this section is shown in [50].

Each process is assigned its own subdomain with ranges in X, Y and Z dimensions between `[myminx,mymaxx]`, `[myminy,mymaxy]` and `[myminz,mymaxz]` respectively. Every process allocates two copies of space each for storing its own subdomain and additionally surrounding cells (so-called "ghost nodes" for which values are computed by other processes but are needed by this process to update values of its boundary cells). Two copies of this space are needed because one acts as a data source for a particular iteration (with values computed in the previous time step) and the other as a target space in the current iteration.

In each iteration of the main loop each cell of the domain is updated. In this application, these updates are performed within function `updatecell(...)`. Code for updates will vary depending on the problem. Similarly, the application uses a `t_cell` data type to represent variables and values assigned to a cell at a given point in time. Again, the meaning of particular variables and corresponding data types are problem dependent.

The application defines new data types for representing walls of cells for which values need to be exchanged between iterations.

In each iteration of the main loop the following steps are performed:

1. Exchange of boundary cell data – this is needed if an implementation starts with initialization of data (before the main loop) which is performed by each process on its subdomain. Communication operations superimposed on subdomains are depicted in Figure 5.10 and processes with communication are shown in Figure 5.11. It should be noted that there are several steps in this communication phase. In each step pairs of processes communicate with each other in parallel which is shown for communication along the X axis.

2. Computations – each process updates values in its subdomain, as follows:

```
for(z=myminz;z<=mymaxz;z++)
  for(y=myminy;y<=mymaxy;y++)
    for(x=myminx;x<=mymaxx;x++)
      updatecell(datanext,data,x,y,z);
```

3. Substitution of pointers `data` and `datanext` for the next iteration.

The code can be compiled and run as follows:

```
mpicc <flags> SPMD-MPI-1.c
mpirun -np 16 ./a.out 400 400 400 100
```

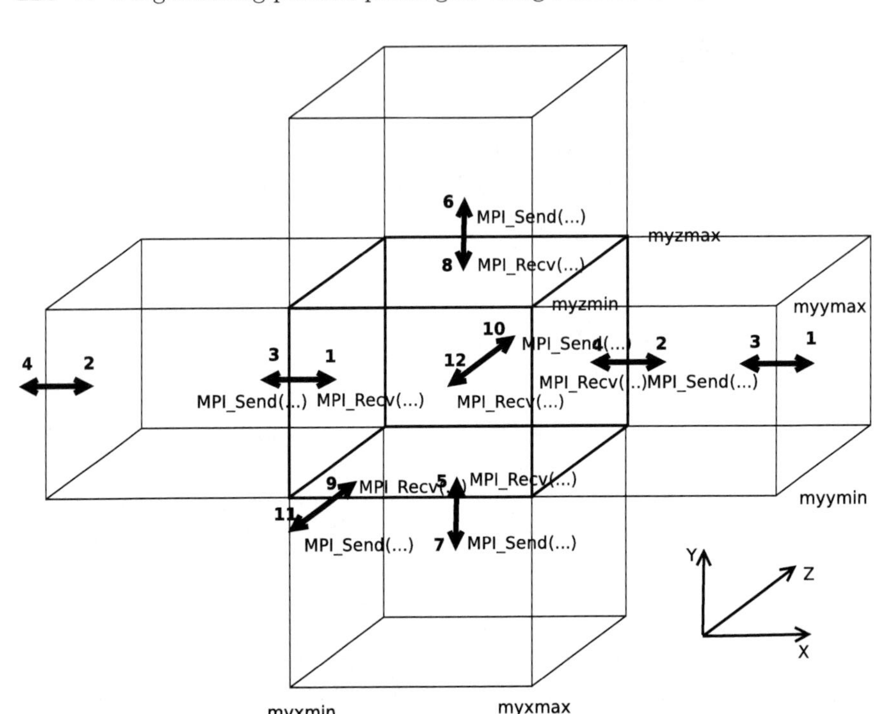

FIGURE 5.10 Subdomain and order of communication operations for an MPI SPMD application

Figure 5.12 presents execution times of the application for various sizes of the domain and 100 iterations and for various numbers of processes on a workstation with 2 x Intel Xeon E5-2620v4, 128 GB RAM. Presented times are best times out of three runs for each configuration.

It should be noted that, depending on a particular problem, the ratio of cell update time compared to necessary communication and synchronization times might vary. In a parallel version this ration is crucial to obtaining good speed-ups.

5.2.2 MPI+OpenMP

The MPI implementation presented in Section 5.2.1 can be extended with support for multithreading using OpenMP. For instance, the following changes can be introduced into the code:

1. Adding `#include <omp.h>`.
2. Adding the following code instead of a call to `MPI_Init`:

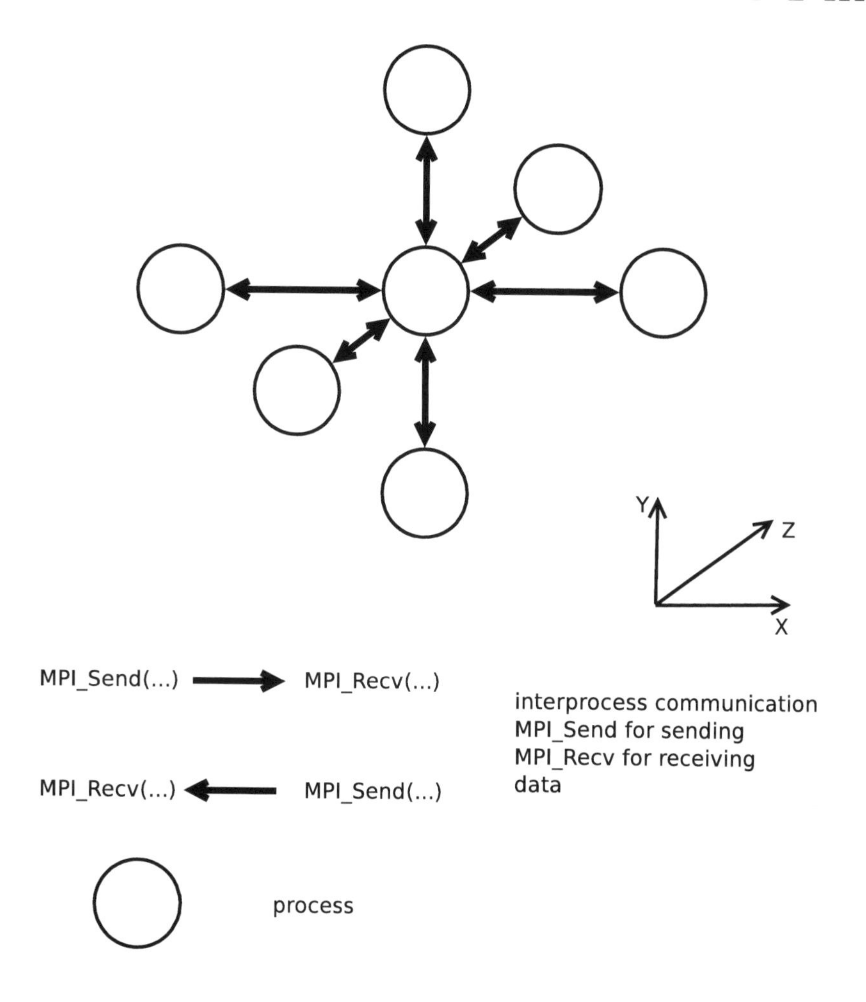

FIGURE 5.11 Processes and communication operations for an MPI SPMD application

```
int threadsupport;

MPI_Init_thread(&argc, &argv,MPI_THREAD_FUNNELED,
&threadsupport);
if (threadsupport<MPI_THREAD_FUNNELED) {
  printf("\nThe implementation does not support
  MPI_THREAD_FUNNELED,
  it supports level %d\n",threadsupport);
  MPI_Finalize();
  return -1;
}
```

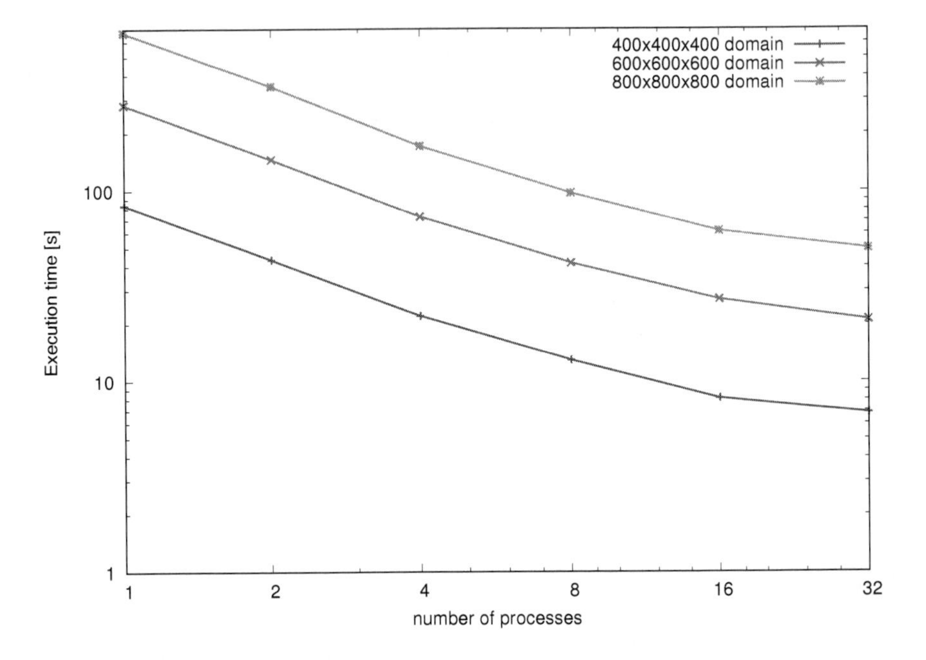

FIGURE 5.12 Execution time of the testbed MPI SPMD application run on a workstation with 2 x Intel Xeon E5-2620v4, 128 GB RAM, 16 physical cores, 32 logical processors

3. Partitioning of loops when updating cells. This can be implemented as follows:

```
#pragma omp parallel for private(x,y) \
shared(myminx,mymaxx,myminy,mymaxy, \
myminz,mymaxz,datanext,data) schedule(static,5)
  for(z=myminz;z<=mymaxz;z++)
    for(y=myminy;y<=mymaxy;y++)
      for(x=myminx;x<=mymaxx;x++)
        updatecell(datanext,data,x,y,z);
```

It should be noted that top level iterations are spread among computing threads statically in blocks of 5 based on the assumption that each cell takes the same amount of time to be updated. Good values for the block size may depend on the value of `mymaxz-myminz+1`. On the other hand, if some cells require more or less time to be computed, dynamic scheduling should be used instead, as described in Section 4.2.2.

Compilation and running in a testbed Linux environment can be performed as follows:

```
mpicc -fopenmp <flags> SPMD-MPI+OpenMP-1.c
export OMP_NUM_THREADS=2
mpirun -np 16 ./a.out 400 400 400 100
```

It should be noted, however, that this implementation starts a parallel region in every iteration of the main loop which results in an additional overhead.

An optimized version of this solution is presented next and is contained in file `SPMD-MPI+OpenMP-2.c`. In this version

```
#pragma omp parallel private(x,y,z,t) \
shared(myminx,mymaxx,myminy,mymaxy, \
myminz,mymaxz,stepcount,datanext,data)
```

is used before the main loop which results in the overhead for parallel region creation only once. Consequently, synchronization is needed within the simulation loop. Important steps within the main loop are executed as follows:

1. Communication by the master thread:

   ```
   #pragma omp master
   {
     <MPI communication>
   }
   ```

2. Synchronization of threads after communication by `#pragma omp barrier`.

3. Parallelization of updates among threads using `#pragma omp for`:

   ```
   #pragma omp for private(x,y) schedule(static,5)
    for(z=myminz;z<=mymaxz;z++)
      for(y=myminy;y<=mymaxy;y++)
        for(x=myminx;x<=mymaxx;x++)
          updatecell(datanext,data,x,y,z);
   ```

   It should be noted that there is an implicit barrier at the end of this code fragment.

4. Substitution of pointers done by the master thread, as follows:

   ```
    #pragma omp master
   {
      if (data==data0) {
   ```

```
        data=data1;
        datanext=data0;
      } else {
        data=data0;
        datanext=data1;
      }
  }
```

Comparison of performance of the two versions is presented in Section 6.3.1.

Figure 5.13 presents execution times of the application for various sizes of the domain and 100 iterations and for various numbers of processes on a workstation with 2 x Intel Xeon E5-2620v4, 128 GB RAM. The number of threads per process was set to 2. Presented times are best times out of three runs for each configuration.

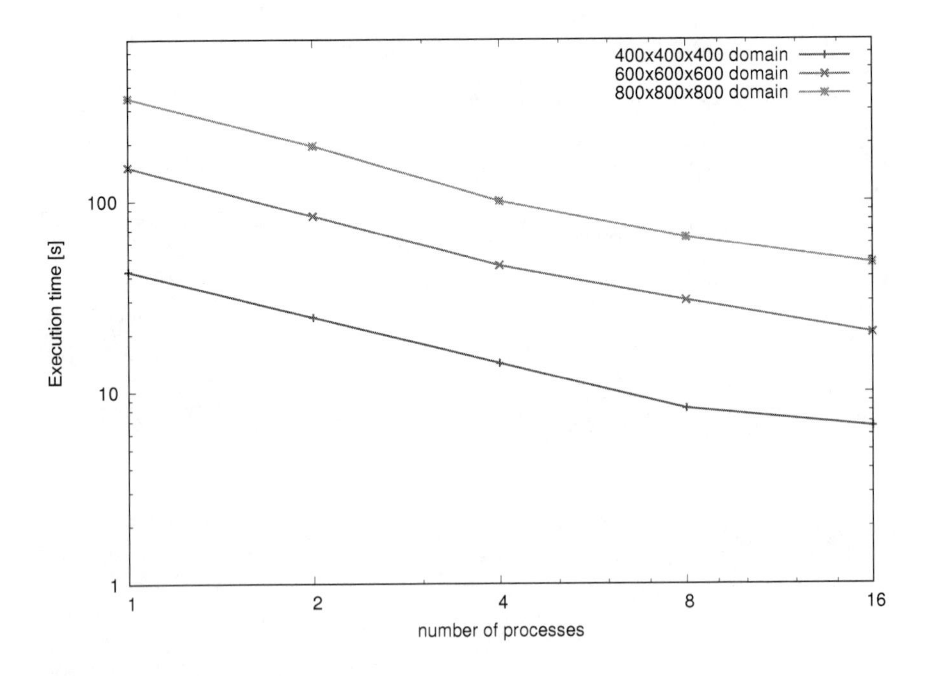

FIGURE 5.13 Execution time of the testbed MPI+OpenMP SPMD application run on a workstation with 2 x Intel Xeon E5-2620v4, 128 GB RAM, 16 physical cores, 32 logical processors, 2 threads per process

5.2.3 OpenMP

Essentially, in a shared memory environment, domain data is accessible to all threads thus only proper synchronization is required between iterations. Consequently, one viable approach could be to use the code of the MPI+OpenMP version presented in Section 5.2.2, simplified to use just 1 process. The MPI part can then be removed altogether leaving just the OpenMP including directives.

5.2.4 MPI+CUDA

The MPI implementation presented in Section 5.2.1 can be extended for use with GPU cards installed within cluster nodes, as shown in Figure 5.14. This is especially interesting when the following are assumed:

1. It is possible to specify the number of GPUs per node in which case an MPI process can control access to a single GPU. While it might be slightly more efficient to use threads, this approach is a straightforward extension of the previous code, in terms of programming.

2. Managed memory is used in which case invocation of a kernel function on a GPU is very similar to an invocation of a function on a CPU. In this case, memory is allocated for an active GPU as indicated in [122].

In general, CUDA-aware MPI implementation functions are able to use GPU buffers which simplifies programming. For instance, it is possible in OpenMPI 1.7+ [88]. According to the documentation [89], OpenMPI properly handles Unified Memory starting from version 1.8.5.

It is assumed that code is run on a cluster, each with one or more GPU cards installed in it. The Unified Memory mechanism allows allocation of space for each subdomain using `cudaMallocManaged(...)` calls, specifically for two copies used interchangeably within the code for successive iterations. Interprocess communication is handled as in the previous non-CUDA versions. No explicit copying between the host and the device is needed thanks to the Unified Memory solution which makes the code very readable and also easy to adapt from a standard MPI version.

The structure of the code is presented in Figure 5.15. Compared to the MPI only code, kernel function `updatecellgpu(...)` defined within a `.cu` is invoked from function `updatecells(...)` which is also included in the same `.cu` file. The latter is called from within a `.c` file. Both the `.cu` and the `.c` files include a header file `.h` with definition of the cell data type. Note that `int` types are used for indices in this version of the code.

Specifically, file `SPMD-MPI+CUDA-2.c` includes:

1. Setting a device.

2. Allocation of memory.

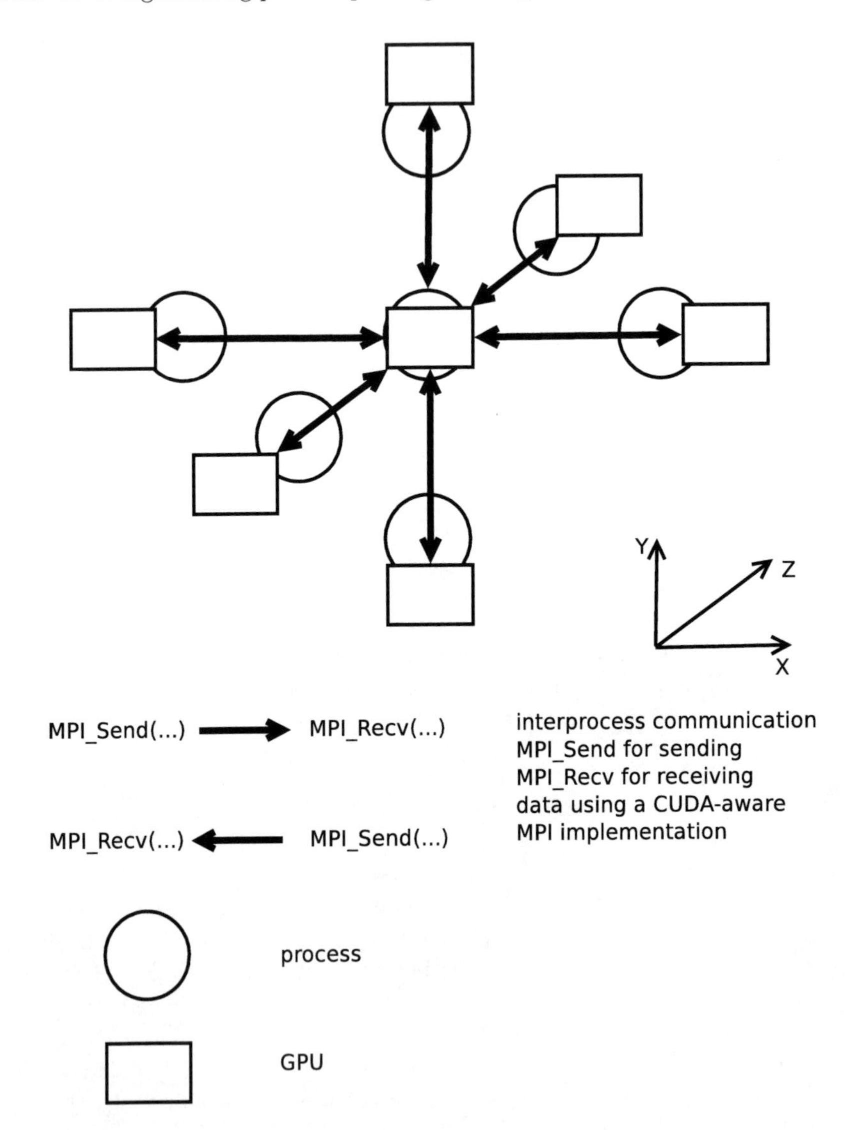

FIGURE 5.14 Architecture of the MPI+CUDA application

3. Data initialization.
4. Type definition.
5. Main iteration loop with the following steps: MPI communication, invocation of function `updatecells(...)` for subdomain update, defined in file `SPMD-MPI+CUDA-2.cu`.
6. Substitution of pointers to source and destination subdomains.

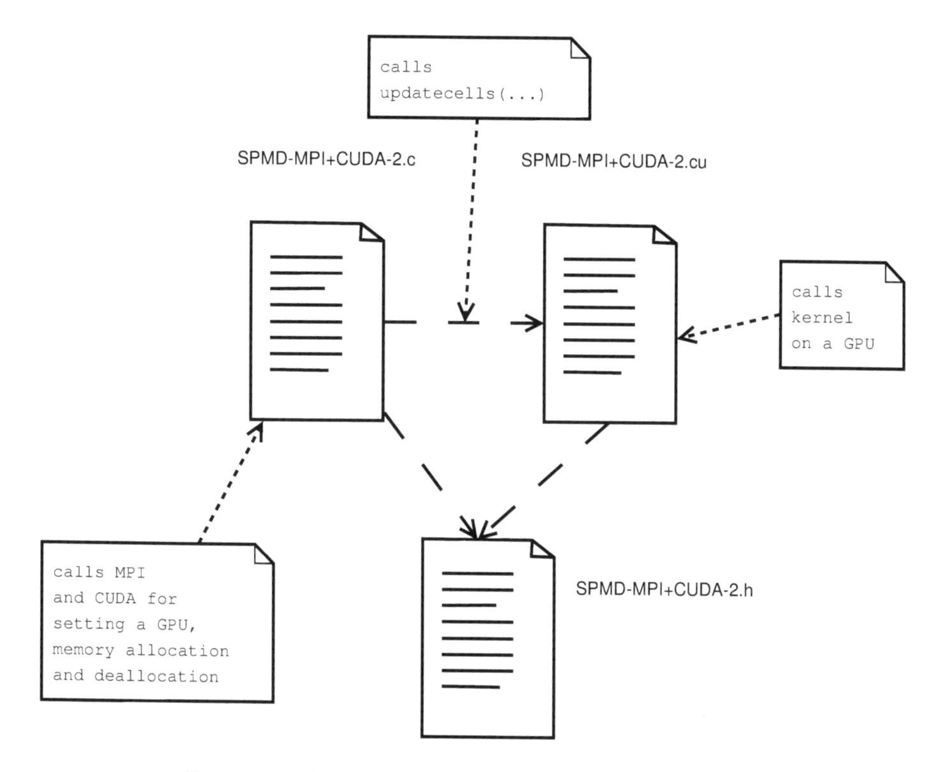

FIGURE 5.15 Source code structure for the MPI+CUDA application

7. Presenting output to the user – there is no need for copying of data to the host by a copy command because Unified Memory allows easy access to results.

8. Freeing memory.

File `SPMD-MPI+CUDA-2.cu` includes, mainly, invocation of a kernel on a GPU in the following function:

```
extern "C" void updatecells(t_cell *datanext,t_cell *data,
int myminx,int mymaxx,int myminy,int mymaxy,int myminz,
int mymaxz,int myblockxsize,int myblockysize,int myblockzsize) {

  // assuming divisible by 16 and 4
  dim3 blocksingrid(((mymaxx-myminx+1)/16),
((mymaxy-myminy+1)/16),((mymaxz-myminz+1)/4));
  dim3 threadsinblock(16,16,4);

  updatecellgpu<<<blocksingrid,threadsinblock>>>(datanext,
data,myminx,myminy,myminz,myblockxsize,myblockysize,
```

```
myblockzsize); // updates the value of the given cell
  if (cudaSuccess!=cudaGetLastError())
    printf("Error during kernel launch in stream");

  cudaDeviceSynchronize();
}
```

The code can be compiled as follows:

```
nvcc -arch=sm_35 -c SPMD-MPI+CUDA-2.cu -o cuda.o
mpicc SPMD-MPI+CUDA-2.c cuda.o -lcudart
```

and subsequently the application can be invoked in the following way, as an example:

```
mpirun -np 2 ./a.out  384 384 960 10 2
```

Results for a test run on a workstation with 2 x Intel Xeon E5-2620v4, 2 x NVIDIA GTX 1070 and 128 GB RAM are shown in Table 5.7.

TABLE 5.7 Execution times [us] for the MPI+CUDA code, workstation with 2 x Intel Xeon E5-2620v4, 2 x NVIDIA GTX 1070 and 128 GB RAM, 16 physical cores, 32 logical processors, application parameters `384 384 960 10 2`

number of processes/GPUs used	Execution time [s]
1	7.174
2	4.668

Additionally, this approach can be further extended in such a way that within each node both GPUs as well as multicore CPUs are utilized. In the presented implementation, though, the following limitations should be considered:

1. It might be difficult to balance load within each node due to the following:
   - the number of cells to be assigned to each GPU may be small and consequently update times may be small compared to the overhead for kernel launch and synchronization,
   - CPU and GPU performance will most likely require dynamic load balancing due to differences in processing speeds.
2. Unified Memory originally did not allow referencing from the host while a GPU is active [122], see a note on Pascal in Section 4.4.7.

The following solutions can be considered:

1. Partitioning of the domain into a number of subdomains equal to the sum of the numbers of cores and numbers of GPUs in the system such that various performances of CPUs and GPUs are considered which results in subdomains of various sizes.

2. Partitioning of the domain into a larger number of subdomains such that subdomains are distributed across computing devices in the system. This is easier to implement starting from the presented code but may result in an overhead for additional process management.

Furthermore, it should be noted that the presented implementation can be improved with overlapping communication and computations, as demonstrated in Section 6.1. This applies to both:

1. internode communication,

2. communication between a host and a GPU such as using streams.

It should also be noted that NVIDIA Multi Process Service (MPS) [119] allows overlapping of kernel execution and memory copy operations from various host processes. This may bring benefits to application execution time and increase utilization of the GPU. Its benefits are shown in Section 6.1.2.

5.3 DIVIDE-AND-CONQUER

5.3.1 OpenMP

OpenMP features nested parallelism (discussed in Section 4.2.5) that allows parallelization at various levels through `#pragma omp parallel` or `#pragma omp sections` directives. However, this may lead to creation of too many threads resulting in performance drop. On the other hand, the total number of threads can be controlled by environment variable `OMP_THREAD_LIMIT`.

It should be noted that a multicore or a manycore type system is preferable in order to make processing of independent subtrees of the divide-and-conquer tree run in parallel. Paper [47] discusses parallelization of divide-and-conquer applications using an OpenMP based framework on an Intel Xeon Phi coprocessor. The paper presents impact of the maximum depth for allowing thread spawning on the execution time. The implementation presented in Listing 5.14 uses OpenMP for parallelization of a divide-and-conquer application. While the solution presented in [47] first generates a number of data packets each of which is then processed and its computations possibly divided and results merged, the solution presented in this book starts with a single input data set which gets partitioned. Specifically, function `integratedivideandconquer(...)` in Listing 5.14 is invoked recursively. Each invocation is to process an input range submitted as input. If a certain depth has been reached then an integrate for this range is computed using a rectangle based method with a certain accuracy. Otherwise, a check is performed whether the range should be partitioned into two subranges or not. This is done by comparing whether rectangle based computations with a certain rectangle width and width/2 differ considerably. If true, another check is performed whether processing is close enough to the root to spawn computations in two threads or not. In both cases computations are started recursively by invoking function `integratedivideandconquer(...)` for the two subranges. It should also be noted that the implementation provided in file `divide-and-conquer-OpenMP-1.c` enables nested parallelism with `omp_set_nested(1)` in function `main()`. This processing scheme is also shown visually in a diagram in Figure 5.16.

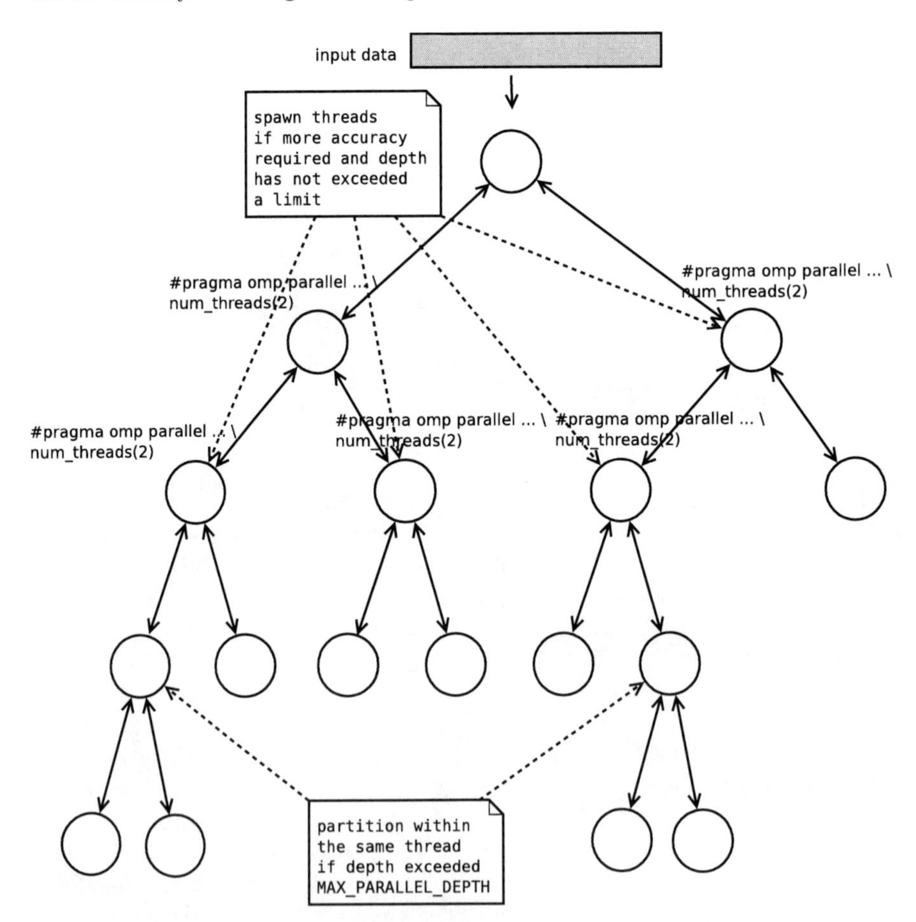

FIGURE 5.16 Divide-and-conquer scheme with OpenMP

Listing 5.14 Divide-and-conquer application for adaptive integration using OpenMP – function integratedivideandconquer(...)

```
double integratedivideandconquer(double start,double end,int
    depth) {

  double step2result=integratesimple(start,end,ACCURACY/2);
  double result;

  result=step2result;

  if (depth==MAX_RECURSION_DEPTH)
    return result;

  // now repeat it with a smaller resolution
    double stepresult=integratesimple(start,end,ACCURACY);

  // check if we need to go deeper at all

  if (fabs(step2result-stepresult)>0.0000000001) {

    // now check whether we are not deep enough to use new
    threads
    if (depth<=MAX_PARALLEL_DEPTH) {

      int threadid,threadcount;
      // run two parts in distinct threads

      result=0;
#pragma omp parallel private(threadid,threadcount) shared(start,
    end,depth) reduction(+:result) num_threads(2)
      {
        threadcount=omp_get_num_threads();
        threadid=omp_get_thread_num();

        if (threadcount==2) { // two threads active -- spawn
    computations
          if (threadid==0) {
            result=integratedivideandconquer(start,(start+end)
    /2,depth+1);
          } else {
            result=integratedivideandconquer((start+end)/2,end,
    depth+1);
          }
        } else { // only one thread active
```

```
        result=integratedivideandconquer(start,(start+end)/2,
depth+1)+
          integratedivideandconquer((start+end)/2,end,depth+1)
;
      }
    }
  } else { // continue partitioning within this thread
    result=integratedivideandconquer(start,(start+end)/2,depth
+1)+
      integratedivideandconquer((start+end)/2,end,depth+1);
  }
}

  return result;
}
```

The code can be compiled as follows:

```
gcc -fopenmp <flags> divide-and-conquer-OpenMP-1.c -lm
```

and run using the `taskset` command to indicate processors on which the multithreaded application can run, as shown in Section 6.4.

Table 5.8 presents execution times for the aforementioned code for various numbers of available cores on a workstation with 2 x Intel Xeon E5-2620v4, 128 GB RAM. Best execution times out of three runs are presented for each configuration. It should be noted that limited speed-up stems from limited possibility of parallelization (a tree where only lower levels can be parallelized) as well as limited sizes of subtrees that can be parallelized.

TABLE 5.8 Execution times [s] for parallel OpenMP divide-and-conquer code

number of threads	execution time [s]
1	42.593
2	26.445
4	22.234
8	20.186
16	20.150
32	19.716

5.3.2 CUDA

CUDA, as discussed in Section 4.4.6, with proper cards and software versions, supports dynamic parallelism that in turn allows launching a kernel from within a kernel. It allows natural mapping of divide-and-conquer applications onto the API taking into account the technological constraints listed in Section 4.4.6.

In this section, adaptive numerical integration of a function over a given range is presented using CUDA dynamic parallelism. Assumptions for this implementation are as follows:

1. An initial (parent) kernel is launched which starts with an original range to be integrated.

2. Each block computes its range using the standard rectangle based method with `step` and `step/2` for a rectangle width. If the difference between these two results is below a given threshold then the latter result is considered to be accurate enough and the threads return. Otherwise the subrange is divided into two smaller subranges for which computations are started recursively using dynamic parallelism. It should be noted that those launches are submitted to two different and newly created streams for potentially increased concurrency, as discussed in [10].

3. Within the kernel a condition is checked whether computations reach a certain depth in which case the current result is assumed to be the final one.

4. The initial, top-level kernel launch starts several blocks. In the nested calls, in this particular approach, two kernel launches are performed by the thread with `threadIdx.x` equal to `0`, each with 1 block.

The idea of the approach is presented in Figure 5.17 while the recursive kernel implementation is presented in Listing 5.15.

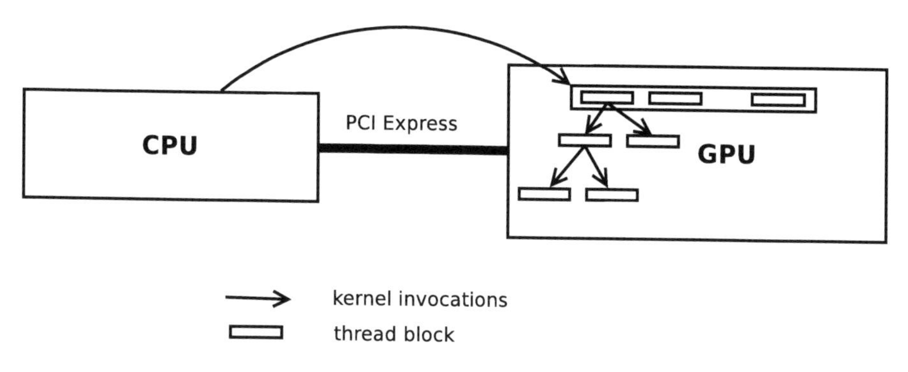

FIGURE 5.17 Divide-and-conquer scheme with CUDA and dynamic parallelism

Listing 5.15 Divide-and-conquer application for adaptive integration using CUDA with dynamic parallelism – kernel implementation

```
__global__
void integratedp(double *result,double starti,double endi,double
    step,int depth) {
  extern __shared__ double sresults[]; // use dynamically
    allocated shared memory
  long counter;
  cudaStream_t stream1,stream2;
  double width=(endi-starti)/(double)gridDim.x;
  double start=starti+blockIdx.x*width;
  double end=start+width;

  sresults[threadIdx.x]=integratesimple(start,end,step/2);
  __syncthreads();
  for(counter=THREADS_PER_BLOCK/2;counter>0;counter/=2) {
    if (threadIdx.x<counter)
      sresults[threadIdx.x]+=sresults[threadIdx.x+counter];
    __syncthreads();
  }

  if (threadIdx.x==0)
    *(result+blockIdx.x)=sresults[0];

  if (depth==MAX_RECURSION_DEPTH)
     return;

  // now repeat it with a smaller resolution
  sresults[threadIdx.x]=integratesimple(start,end,step);
  __syncthreads();
  for(counter=THREADS_PER_BLOCK/2;counter>0;counter/=2) {
    if (threadIdx.x<counter)
      sresults[threadIdx.x]+=sresults[threadIdx.x+counter];
    __syncthreads();
  }

  // now thread 0 checks whether to go deeper
  if (threadIdx.x==0) {
    if (fabs(*(result+blockIdx.x)-sresults[0])>0.0000001) { //
    need to invoke kernel recursively
          cudaStreamCreateWithFlags(&stream1,
    cudaStreamNonBlocking);
        integratedp<<<1,THREADS_PER_BLOCK,THREADS_PER_BLOCK*
    sizeof(double),stream1>>>(result+blockIdx.x,start,(start+end
    )/2,step/2,depth+1);
```

```
        cudaStreamCreateWithFlags(&stream2,
    cudaStreamNonBlocking);
        integratedp<<<1,THREADS_PER_BLOCK,THREADS_PER_BLOCK*
    sizeof(double),stream2>>>(result+blockIdx.x+(long)BLOCKS
    *(1<<(MAX_RECURSION_DEPTH-depth-1)),(start+end)/2,end,step
    /2,depth+1);
    } else *(result+blockIdx.x+(long)BLOCKS*(1<<(
    MAX_RECURSION_DEPTH-depth-1)))=0;
  }
  __syncthreads();
  if (threadIdx.x==0) {
    cudaDeviceSynchronize();
    cudaStreamDestroy(stream1);
    cudaStreamDestroy(stream2);
  }
  __syncthreads();
  if (threadIdx.x==0)
    *(result+blockIdx.x)+=*(result+blockIdx.x+(long)BLOCKS*(1<<(
    MAX_RECURSION_DEPTH-depth-1)));
}
```

The code can be compiled and run as follows:

```
nvcc -arch=sm_35 -rdc=true \
divide-and-conquer-CUDA-dynamic-parallelism1.cu \
-L /usr/lib/x86_64-linux-gnu -lcudadevrt
./a.out
```

```
The final result is 13.122364
```

This particular code was tested on a mobile computer with Intel Core i5-7200U CPU and NVIDIA GeForce 940MX GPU.

Note that accuracy of the result may vary depending on the values within the algorithm such as the adaptive threshold.

5.3.3 MPI

For MPI, two implementations are proposed and discussed:

1. A balanced implementation in which an input problem (and consequently data) is divided into problems of computationally equal sizes. This is repeated recursively until a certain depth of the divide-and-conquer tree is reached and allows structuring computations and communication well. It also requires the number of processes equal to the number of leaves generated. There are several algorithms that might benefit from such an approach, for example:

- mergesort which is a balanced algorithm,
- partitioning an image (presumably large) into parts on which various processes might apply filters in parallel,
- numerical algorithms including matrix multiplication, rectangle based numerical integration etc.

2. A flexible approach with requesting and fetching work dynamically. In this case, it is assumed that the number of subproblems in various nodes of the divide-and-conquer tree might differ. Similarly, a computational effort and consequently execution times for processing various leaves might differ, also when executed on identical CPUs or cores. In general, such a tree might only be known when the algorithm runs and thus cannot be predicted in advance, especially regarding execution times of particular subtrees.

5.3.3.1 Balanced version

The first code assumes type `t_datapacket` that encapsulates data on which the algorithm works. An implementation of numerical integration of a function defined in function `fcpu(...)` over a given range is discussed.

It distinguishes the following functions, applicable to the divide-and-conquer scheme:

(a) `generateinputdata(...)` for generation input data based on program arguments,

(b) `partition(...)` for partitioning input data into a certain number of subproblems represented by data packets,

(c) `computationslocal(...)` for computations in leaves of the divide-and-conquer tree,

(d) `merge(...)` for merging a given number of subresults into an output data packet.

Figure 5.18 presents the working scheme using MPI with process ranks, functions and MPI calls. Processes are arranged into a binary tree structure in such a way that in every iteration, pairs of processes communicate and parent processes pass one of subproblems to another process. This is continued until all processes receive their parts. Then processes process data associated with their subproblems by invoking function `computationslocal(...)` in parallel. What follows is parallel integration of results from each process which is done inversely to partitioning in a similar tree like structure in which pairs of processes communicate in every iteration.

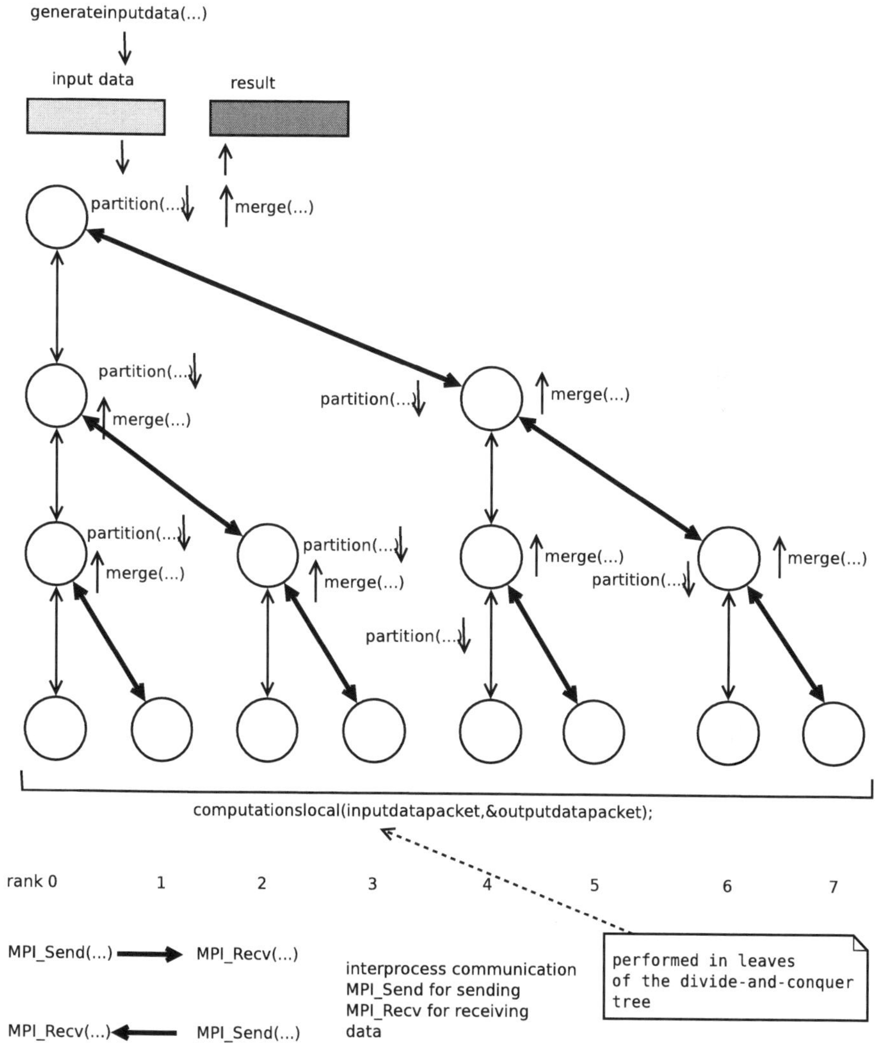

FIGURE 5.18 Balanced divide-and-conquer scheme with MPI

Listing 5.16 presents MPI code for distribution of input data among MPI processes while Listing 5.17 presents code for merging partial results from processes, assuming more than 1 process.

Listing 5.16 Basic divide-and-conquer application for balanced computations using MPI – partitioning

```
if (!myrank) {
```

```
    partition(inputdatapacket,&outputdatapackets,&
      outputdatapacketscount);

    // free the current data packet
    freedata(inputdatapacket);

    // assuming 2 output data packets for this application
    // now the first data packet will be mine
    inputdatapacket=&(outputdatapackets[0]);
    // and send the other data packet to a following process
    MPI_Send(outputdatapackets[1].elements,2,MPI_DOUBLE,
      proccount/2,0,MPI_COMM_WORLD);
  }

  int alreadyreceived=0;
  for(int currentskip=proccount/2;currentskip>=1;currentskip
      /=2) {
    if (!(myrank%currentskip)) { // then I am involved in the
        given step

      // receive if possible
      if ((myrank-currentskip>=0) && (!alreadyreceived)) {
        MPI_Recv(inputdatapacket->elements,2,MPI_DOUBLE,
      myrank-currentskip,0,MPI_COMM_WORLD,MPI_STATUS_IGNORE);
        alreadyreceived=1;
      }

      // now if it is not the last step then the input range
      needs to be partitioned
      if (currentskip/2>=1) { // partition
        partition(inputdatapacket,&outputdatapackets,&
      outputdatapacketscount);

        // free the current data packet
        freedata(inputdatapacket);
        // assuming 2 output data packets for this
      application
        // now the first data packet will be mine
        inputdatapacket=&(outputdatapackets[0]);
        // and send the other data packet to a following
      process

        MPI_Send(outputdatapackets[1].elements,2,MPI_DOUBLE,
      myrank+currentskip/2,0,MPI_COMM_WORLD);
```

```
    }
  }
}
```

Listing 5.17 Basic divide-and-conquer application for balanced computations using MPI – merging

```
if (myrank%2) // then send the data to process with rank
    myrank-1
  MPI_Send(outputdatapacket->elements,1,MPI_DOUBLE,myrank
    -1,0,MPI_COMM_WORLD);

for(int currentskip=2;currentskip<=proccount;currentskip
    *=2) {
  if (!(myrank%currentskip)) { // then I (process) am
    involved in the given step

    // processes that partitioned before will now merge
    // these processes have outputdatapackets spaces
    // so it is possible to receive to those spaces
    outputdatapackets[0].elemcount=1;
    outputdatapackets[1].elemcount=1;
    outputdatapackets[0].elements[0]=outputdatapacket->
    elements[0]; // copy my result

    MPI_Recv(outputdatapackets[1].elements,1,MPI_DOUBLE,
    myrank+currentskip/2,0,MPI_COMM_WORLD,MPI_STATUS_IGNORE
    );

    // free the previous data packet
    freedata(outputdatapacket);
    // now merge data
    merge(outputdatapackets,2,&outputdatapacket);
    // result is now in outputdatapacket again
    // and send to an upper process if it is not the last
    iteration
    if (currentskip*2<=proccount) { // if not the last
    iteration
      if (myrank%(currentskip*2)) // then I should send the
     data to process with rank myrank-currentskip
        MPI_Send(outputdatapacket->elements,1,MPI_DOUBLE,
    myrank-currentskip,0,MPI_COMM_WORLD);
    }
  }
}
```

Compilation and running the balanced code can be performed as follows, as an example with the `0 1000` arguments:

```
mpicc <flags> DAC-MPI.c
mpirun -np 16 ./a.out 0 1000
```

Figure 5.19 presents execution times for the code obtained on a workstation with 2 x Intel Xeon E5-2620v4 and 128 GB RAM. Best out of three runs for each configuration are reported.

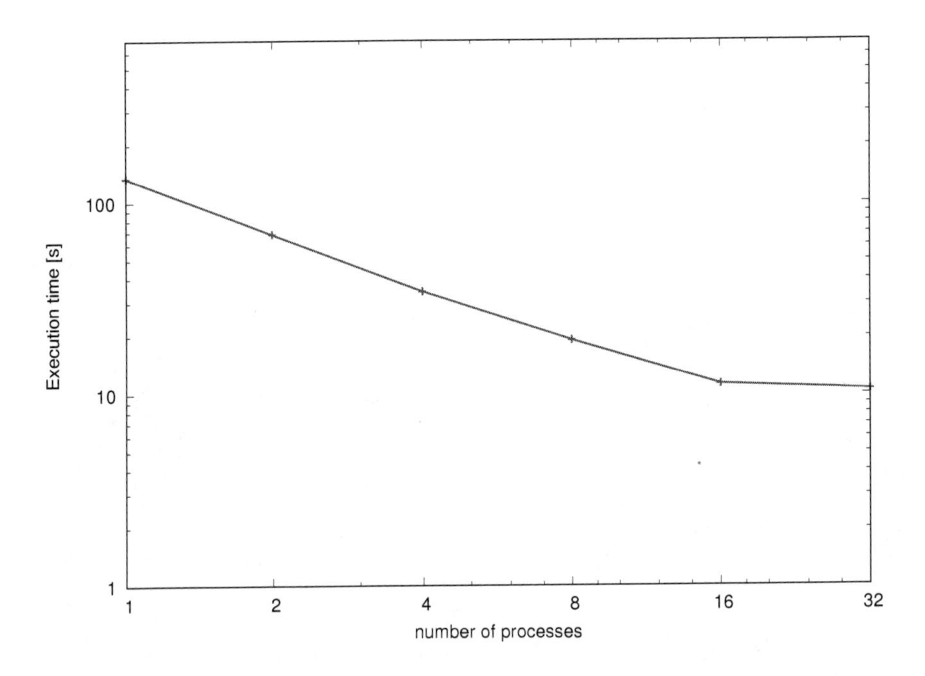

FIGURE 5.19 Execution time of the MPI implementation of the divide and conquer scheme run on a workstation with 2 x Intel Xeon E5-2620v4, 128 GB RAM, 16 physical cores, 32 logical processors

5.3.3.2 *Version with dynamic process creation*

In order to make the most of an available computing infrastructure, load balancing needs to be adopted. There are at least two ways in which such an implementation can be realized:

(a) idle processes requesting work from other processes, the latter cutting off parts of the divide-and-conquer tree and sending for processing, as suggested in Section 3.5 and [38],

(b) a master-slave scheme in which the master would partition the tree (presumably into the number of subtrees considerably larger than the number of processes) while other processes (slaves) would fetch and process those subtrees.

Furthermore, it should be noted that because execution times of subtrees are not known in advance, it is quite probable that mapping one process/thread per core would result in idle times. Even more so because detecting an (partially) idle computing device and arranging new work might consume some time. An alternative approach would be to map more processes/threads per CPU/core in such a way that finishing work by some of the former and potential idle time would be hidden and eliminated by other active processes/threads. This would, however, result in more overhead for process/thread creation, management and synchronization.

A C + MPI code that is able to cope with irregular divide-and-conquer problems is proposed and discussed next. Assumptions for this solution, presented in Figure 5.20, are as follows:

1. Parallelization of processing through dynamic launching new processes handling subtrees up to a predefined level from the root is allowed. Whether a new process is spawned or a subtree is processed within the same process can be determined at runtime e.g. based on current loads of various nodes in a cluster. Similarly, the decision where a new process should be spawned can be made dynamically (Figure 5.21).

2. The number of child nodes is only known at runtime and is returned by function `partition(...)`.

3. Processing nodes at a given level can be performed in any order. It should be noted that in some applications, such as alpha beta search, the order does matter in terms of execution time.

Additionally, setting the maximum level until which spawning is allowed, does impact performance in the following way:

- a larger level (further from the root) allows partitioning of computations into a larger number of parts which makes it easier to balance load across a cluster,
- too large a level adds overhead for additional processes, synchronization and communication between processes.

Figure 5.22 presents files of a proposed implementation, in particular:

– `DAC-MPI-irregular-dynamic-processes-common.c` – includes functions such as data allocation, initialization, partitioning, merging functions and the most important recursive `divideandconquer(...)` function

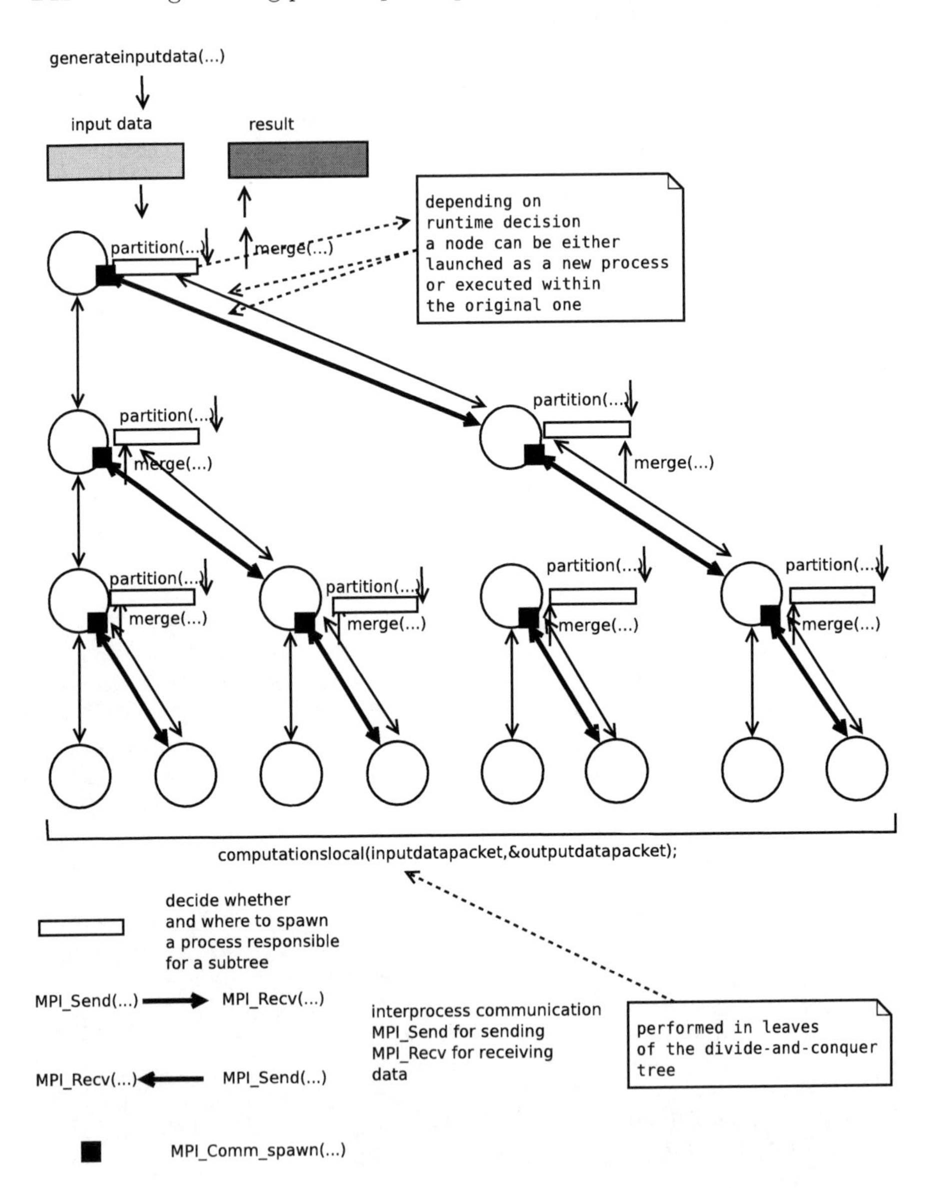

FIGURE 5.20 Divide-and-conquer scheme with MPI and dynamic process creation

shown in Listing 5.18. The latter follows processing of successive nodes of the divide-and-conquer tree. For each node, if the node is a non-leaf node, after partitioning the following decisions need to be performed and can be made at runtime:

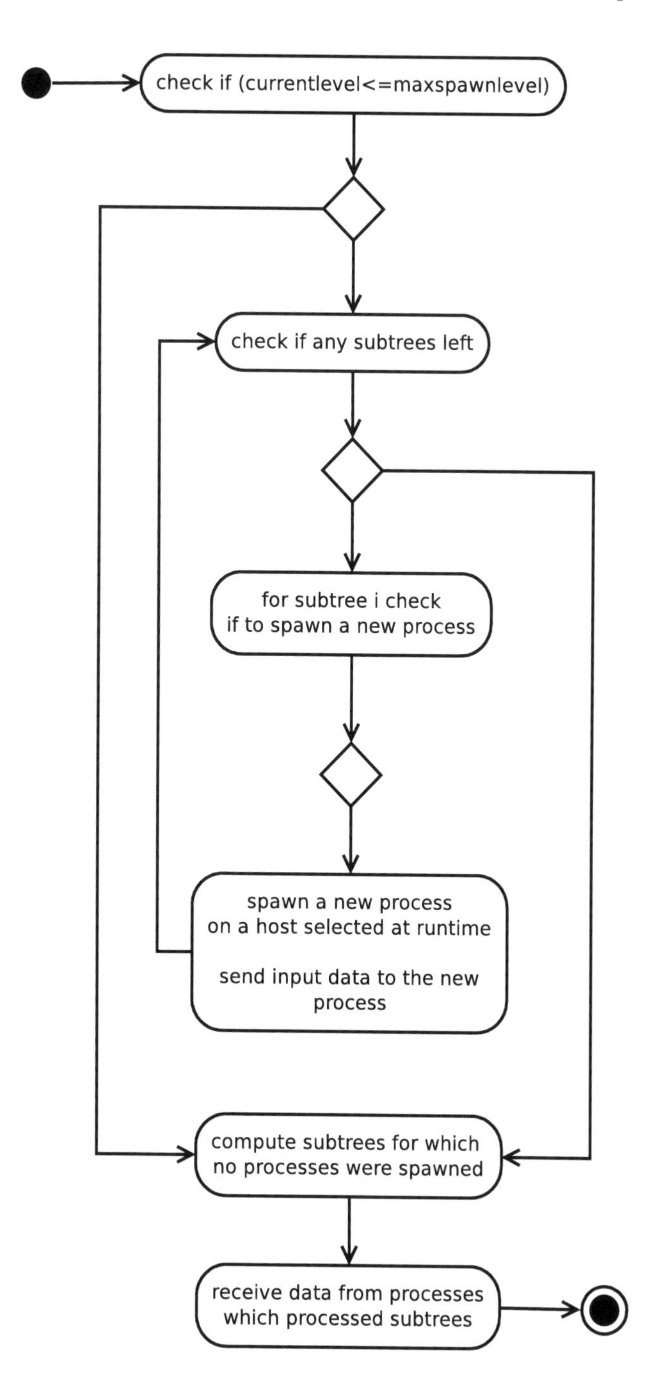

FIGURE 5.21 Handling spawn decisions for divide-and-conquer scheme with MPI and dynamic process creation

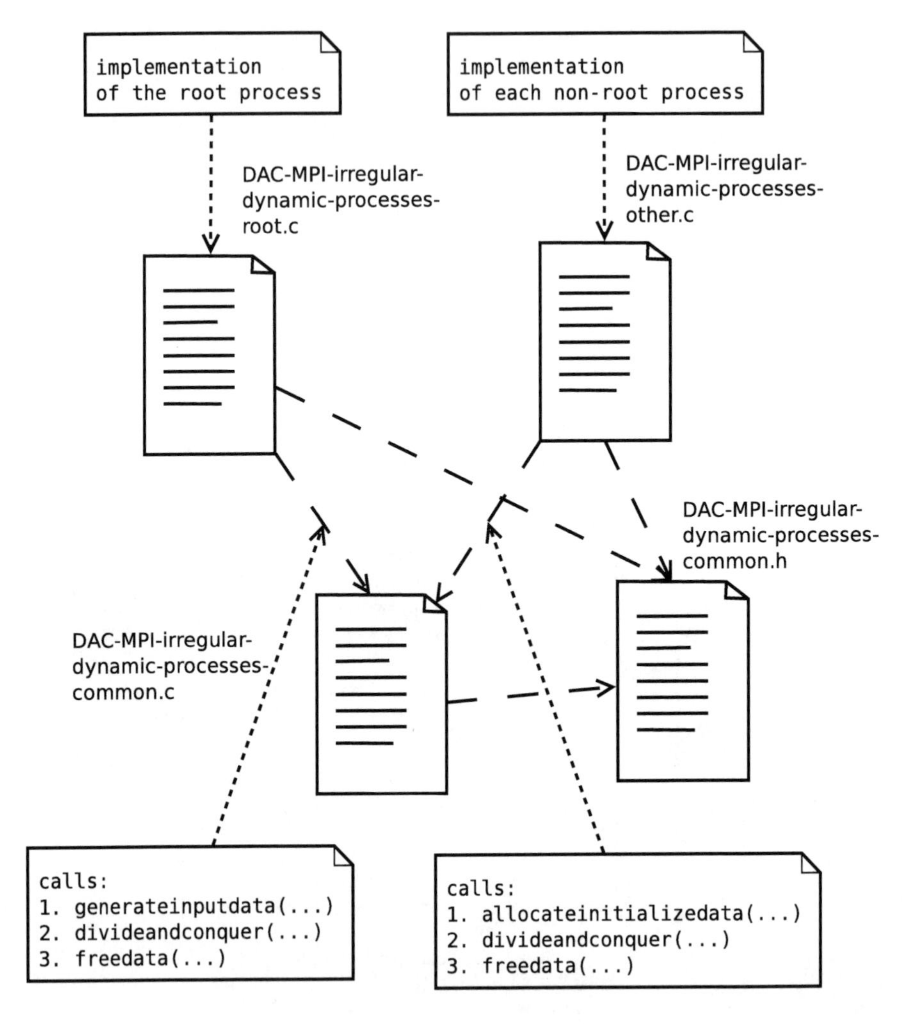

FIGURE 5.22 Source code structure for the divide-and-conquer MPI application with dynamic process creation

- Whether a new process should be spawned for processing of a subtree. A separate function is dedicated to making this decision:

```
int decidespawn(t_datapacket *datapacket,int index) {
  ...
}
```

For instance, based on the index of a subtree it can be decided that the left subtree is launched within the process while the other ones can be spawned as new processes. Furthermore, this function could incorporate information about loads of cluster nodes into the decision so that no new processes are launched if all nodes and cores are already busy processing other subtrees.

- If a new process is to be spawned – on which node it should be started. There is a separate function for this purpose:

  ```
  void selecthosttorun(MPI_Info *info) {
    ...
  }
  ```

  Similarly to the decision on whether to spawn a process, decision on where a process should be started can depend on loads of nodes in the system. For instance, the following instructs to launch a process on host `<hostname>`:

  ```
  MPI_Info_set(*info,"host","<hostname>");
  ```

This scheme is very flexible and can be adjusted depending on particular needs and the state of the system.

Listing 5.18 Divide-and-conquer application using dynamic process creation in MPI – `divideandconquer` function

```
void divideandconquer(t_datapacket *inputdatapacket,
    t_datapacket **result,int currentlevel,int maxlevel,int
     maxspawnlevel) {
  t_datapacket *outputdatapacket;
  t_datapacket *outputdatapackets;
  int outputdatapacketscount;
  int i;

  if (currentlevel==maxlevel) {
    computationslocal(inputdatapacket,result);
    return;
  }
  // otherwise partition and process the tree further

  partition(inputdatapacket,&outputdatapackets,&
    outputdatapacketscount);

  // check if spawn new processes
  if (currentlevel<=maxspawnlevel) {
    MPI_Comm *intercommunicators;
    char *spawnedforpacket; // whether there was a spawn
    for a given data packet
```

```
spawnedforpacket=(char *)malloc(outputdatapacketscount*
sizeof(char));
if (!spawnedforpacket)
  errorexit("Not enough memory");

intercommunicators=(MPI_Comm *)malloc(
outputdatapacketscount*sizeof(MPI_Comm));
if (!intercommunicators)
  errorexit("Not enough memory");

// launch processes for processing of other data
packets
int errorcode;
// use arguments to pass the level
char *argv[2];
argv[0]=(char *)malloc(sizeof(char)*10);
if (!(argv[0]))
  errorexit("Not enough memory");
snprintf(argv[0],10,"%d",(currentlevel+1));
argv[1]=NULL;

// here it is possible to decide whether to spawn
processes immediately or wait until load on
// nodes drops below a threshold and where to spawn
processes
// e.g. do it one by one and check if the number of
active processes is higher or lower than a threshold

for(i=0;i<outputdatapacketscount;i++) {
  if (spawnedforpacket[i]=decidespawn(&(
outputdatapackets[i]),i)) {
    MPI_Info info;
    MPI_Info_create(&info);
    // invoke a function that would return the host to
run a process on (this function may wait
    // until load drops below a threshold)
    selecthosttorun(&info);
    MPI_Comm_spawn("DAC-MPI-irregular-dynamic-processes
-other",argv,1,info,0, MPI_COMM_SELF,&(
intercommunicators[i]),&errorcode);
    if (errorcode!=MPI_SUCCESS)
      errorexit("Error spawning a child process.");
    // and send the other data packets to following
process(es)
```

```
      MPI_Send(outputdatapackets[i].elements,2,MPI_DOUBLE
  ,0,0,intercommunicators[i]);
    }
  }

  for(i=0;i<outputdatapacketscount;i++)
    if (!(spawnedforpacket[i])) {  // process these
  within this process
      inputdatapacket=&(outputdatapackets[i]);
      // go lower with my data packet
      divideandconquer(inputdatapacket,&outputdatapacket,
  currentlevel+1,maxlevel,maxspawnlevel);
      // now my result is pointed by outputdatapacket
      copydatapacket(outputdatapacket,&(outputdatapackets
  [i]));
      // free the previous data packet
      freedata(outputdatapacket);
    }

  // and then gather results from the other processes
  for(i=0;i<outputdatapacketscount;i++)
    if (spawnedforpacket[i]) {
      outputdatapackets[i].elemcount=1; // for this
  application, can be more
      // note that order here might be any but we need to
   wait for results anyway
      MPI_Recv(outputdatapackets[i].elements,1,MPI_DOUBLE
  ,i-1,0,intercommunicators[i],MPI_STATUS_IGNORE);

      MPI_Comm_disconnect(&(intercommunicators[i]));
  }

  // now merge data
  merge(outputdatapackets,outputdatapacketscount,&
  outputdatapacket);
  free(spawnedforpacket);
} else { // process the tree within this process
  t_datapacket *outputdatapacketslowerlevel=
  allocatedatapackets(outputdatapacketscount,1);
  t_datapacket *outputdatapackettemp;
  for(i=0;i<outputdatapacketscount;i++) {
    divideandconquer(&(outputdatapackets[i]),&
  outputdatapackettemp,currentlevel+1,maxlevel,
  maxspawnlevel);
```

```
      copydatapacket(outputdatapackettemp,&(
    outputdatapacketslowerlevel[i]));
    }
    freedata(outputdatapackets); // not needed anymore
    // now merge results
    merge(outputdatapacketslowerlevel,
    outputdatapacketscount,&outputdatapacket);
    freedatapackets(outputdatapacketslowerlevel,
    outputdatapacketscount);
  }
  *result=outputdatapacket;
}
```

– `DAC-MPI-irregular-dynamic-processes-root.c` – includes code for the top level root process that calls the following functions: `generateinputdata(...)` for generation of input data, top level invocation of `divideandconquer(...)` that starts processing of the whole tree and freeing previously allocated data. Key lines of this code are shown in Listing 5.19.

Listing 5.19 Divide-and-conquer application using dynamic process creation in MPI – key code lines of the root process

```
MPI_Init(&argc,&argv);

// check command line parameters
if (argc<3)
  errorexit("\nSyntax: dac-dynamic <startofrange> <
    endofrange>");
MPI_Comm_size(MPI_COMM_WORLD,&proccount);

if (proccount!=1)
  errorexit("\nThe must be only 1 root process.\n");

inputdatapacket=generateinputdata(argc,argv);
divideandconquer(inputdatapacket,&outputdatapacket,1,10,4);
// display the result
printf("\nResult is %f\n",outputdatapacket->elements[0]);

freedata(inputdatapacket);
freedata(outputdatapacket);
MPI_Finalize();
```

– `DAC-MPI-irregular-dynamic-processes-other.c` – includes code for non-root processes spawned using `MPI_Comm_spawn(...)`. This code includes receiving input data from the parent using `MPI_Recv(...)`, allocating and initializing data using `allocateinitializedata(...)`, launching computations for a subtree with `divideandconquer(...)` and the level received upon start, sending out the result with `MPI_Send(...)` and freeing previously allocated data. Key lines of this code are shown in Listing 5.20.

Listing 5.20 Divide-and-conquer application using dynamic process creation in MPI – key code lines of non-root processes

```
MPI_Init(&argc,&argv);
// check command line parameters
if (argc<2)
  errorexit("\nSyntax: dac-dynamic-other <level>");
level=atoi(argv[1]);

MPI_Comm_get_parent(&parentcommunicator);
if (parentcommunicator==MPI_COMM_NULL)
  errorexit("No parent communicator.");

// receive input from the parent
double input[2];
MPI_Recv(input,2,MPI_DOUBLE,0,0,parentcommunicator,
    MPI_STATUS_IGNORE);

// now initialize an input data packet using the received
    data
inputdatapacket=allocateinitializedata(input,2);

divideandconquer(inputdatapacket,&outputdatapacket,level
    ,10,4);

// now send the result back to the parent
MPI_Send(outputdatapacket->elements,1,MPI_DOUBLE,0,0,
    parentcommunicator);

MPI_Comm_disconnect(&parentcommunicator);

freedata(inputdatapacket);
freedata(outputdatapacket);
MPI_Finalize();
```

The following tests were performed on a workstation with 2 x Intel Xeon E5-2620v4 and 128 GB RAM. The code can be compiled as follows:

```
mpicc -c DAC-MPI-irregular-dynamic-processes-common.c \
-o DAC-MPI-irregular-dynamic-processes-common.o
mpicc DAC-MPI-irregular-dynamic-processes-root.c \
DAC-MPI-irregular-dynamic-processes-common.o \
-o DAC-MPI-irregular-dynamic-processes-root
mpicc DAC-MPI-irregular-dynamic-processes-other.c \
DAC-MPI-irregular-dynamic-processes-common.o \
-o DAC-MPI-irregular-dynamic-processes-other
```

and run in the following way (hostfile contains a list of hosts):

```
mpirun -hostfile ./hostfile -np 1 \
./DAC-MPI-irregular-dynamic-processes-root 0 100

Result is 4.615121
mpirun -hostfile ./hostfile -np 1 \
./DAC-MPI-irregular-dynamic-processes-root 0 1000

Result is 6.908755
```

CHAPTER 6

Optimization techniques and best practices for parallel codes

CONTENTS

6.1 DATA PREFETCHING, COMMUNICATION AND COMPUTATIONS OVERLAPPING AND INCREASING COMPUTATION EFFICIENCY

Efficient parallelization of application execution should lead to minimization of application execution time. This generally requires the following:

1. Engaging all compute units for computations that can be achieved through load balancing in order to avoid idle times on the compute units.
2. Minimization of communication and synchronization that result from parallelization.

In some instances, engaging a compute unit may result in unwanted idle times, specifically (shown in Figure 3.6a):

1. Since computation of a data packet requires prior sending or provision of the data packet, idle time may occur before computations.
2. Similarly, after a data packet has been computed, then a process or a thread will usually request another data packet for processing.

Such idle times may show up at various levels in a parallel system, including:

1. Communication between nodes in a cluster when a process requests a data chunk from another process, especially running on a different node.
2. Communication between a host and a GPU within a single node since a thread running on a GPU will need input data initially stored in the host RAM.
3. Fetching input data from global memory on a GPU. Specifically, using shared memory and registers within a GPU is much faster than fetching data from global memory. As a consequence, fetching input data from global memory may become a bottleneck.

A solution to this problem is to prefetch data before it is actually processed. Specifically, at the same time when a certain data packet is processed, another can be fetched in the background such that it is already available when processing of the former data packet has finished. This universal approach can be implemented using various APIs. Specific pseudocodes with proper API calls are provided next.

6.1.1 MPI

There are at least two programming approaches in MPI that allow implementation of overlapping communication and computations and data prefetching. Listing 6.1 presents an approach with non-blocking API calls described in detail in Section 4.1.11. The solution uses `MPI_I*` calls for starting fetching data. Since these are non-blocking calls, a calling process can perform computations immediately after the call. The latter only issues a request for starting communication. After computations, i.e. processing of a data packet, have completed, non-blocking communication needs to be finalized using `MPI_Wait` and processing of the just received data packet can follow. If there are to be more data packets processed then a new data packet can be fetched before computations start.

Listing 6.1 Receiving data using MPI and overlapping communication and computations with non-blocking calls

```
// the first data packet can be received using MPI_Recv

MPI_Recv(inputbuffer,...);
packet=unpack(inputbuffer);

while (shallprocess(packet)) {
  // first start receiving a data packet
  MPI_Irecv(inputbuffer,...,&mpirequest);

  // process the already available data packet
  process(&packet);

  // now finish waiting for the next data packet
  MPI_Wait(&mpirequest);

  // unpack data so that the buffer can be reused
  packet=unpack(inputbuffer);

}
...
```

Alternatively, the code without unpacking of data from a buffer and using two buffers instead is shown in Listing 6.2.

Listing 6.2 Receiving data using MPI and overlapping communication and computations with non-blocking calls and using two buffers

```
// the first data packet can be received using MPI_Recv
buffer=inputbuffer0;
MPI_Recv(buffer,...);

while (shallprocess(buffer)) {

  if (buffer==inputbuffer1) {
    buffer=inputbuffer0;
    prevbuffer=inputbuffer1;
  } else {
    buffer=inputbuffer1;
    prevbuffer=inputbuffer0;
  }

  // first start receiving a data packet
  MPI_Irecv(buffer,...,&mpirequest);

  // process the already available data packet
  process(prevbuffer);

  // now finish waiting for the next data packet
  MPI_Wait(&mpirequest);

}
...
```

In fact, a slave process would normally send back its results to the parent process. Overlapping sends and processing of subsequent data, or packets can also be arranged. Such a solution is shown in Listing 6.3.

Listing 6.3 Receiving data and sending results using MPI and overlapping communication and computations with non-blocking calls and using two buffers

```
MPI_Request requests[2]; // two requests:
// requests[0] used for receive
// requests[1] used for send

requests[1]=MPI_REQUEST_NULL; // do not consider
// the request for send in the first iteration

// the first data packet can be received using MPI_Recv
buffer=inputbuffer0;
MPI_Recv(buffer,...);
```

```
while (shallprocess(buffer)) {

  if (buffer==inputbuffer1) {
    buffer=inputbuffer0;
    prevbuffer=inputbuffer1;
    prevresultbuffer=outputbuffer1;
  } else {
    buffer=inputbuffer1;
    prevbuffer=inputbuffer0;
    prevresultbuffer=outputbuffer0;
  }

  // first start receiving a data packet
  MPI_Irecv(buffer,...,&(requests[0]));

  // process the already available data packet
  process(prevbuffer,prevresultbuffer);

  // now finish waiting for the next data packet
  MPI_Waitall(2,requests,statuses); // note that in the first
    iteration
  // only requests[0] will be active

  // now start sending back the result
  MPI_Isend(prevresultbuffer,...,&(requests[1]));

}

...
```

Another programming approach in MPI can involve two threads in a process, each for performing a distinct task. This, however, requires an MPI implementation that supports multithreading in the required mode (see Section 4.1.14). Then the approach could be as follows – threads would perform their own tasks:

1. Communication i.e. fetching input data and possibly sending results; there can also be a separate thread for sending out results.
2. Processing the already received data packet(s).

This approach is natural and straightforward but requires proper synchronization among the threads. Such synchronization can be implemented e.g. using Pthreads [129, Chapter 4]. The most natural way of implementing such a solution using three threads (receiving, processing and sending) would be to use two queues for incoming data packets and outgoing packets with results, as shown in Section 5.1.4.

6.1.2 CUDA

Implementation of overlapping communication between the host and a GPU and processing on the GPU can be done using streams described in Section 4.4.5. Specifically, using two streams potentially allows overlapping communication between page-locked host memory and a device (in one stream), computations on the device (launched in another stream) as well as processing on the host:

```
cudaStreamCreate(&streamS1);
cudaStreamCreate(&streamS2);

cudaMemcpyAsync(devicebuffer1,sourcebuffer1,copysize1,
cudaMemcpyHostToDevice,streamS1);
kernelA<<<gridsizeA,blocksizeA,0,streamS1>>>(...);
cudaMemcpyAsync(hostresultbuffer1,
deviceresultbuffer1,copyresultsize1,
cudaMemcpyDeviceToHost,streamS1);

cudaMemcpyAsync(devicebuffer2,sourcebuffer2,copysize2,
cudaMemcpyHostToDevice,streamS2);
kernelB<<<gridsizeB,blocksizeB,0,streamS2>>>(...);
cudaMemcpyAsync(hostresultbuffer2,
deviceresultbuffer2,copyresultsize2,
cudaMemcpyDeviceToHost,streamS2);

processdataonCPU();

cudaDeviceSynchronize();
```

As indicated in [137], the issue order of commands may have an impact on execution time. This is due to the fact that host to device, device to host and kernel launch commands are added to respective queues on the device based on the issue order. Commands in a queue in a given stream may need to wait for a sequence of commands issued to another stream if they are preceded in a queue on the device.

Additionally, when using CUDA, the Multi-Process Service (MPS) can be used to increase performance in some scenarios for GPUs with Hyper-Q. Specifically, MPS provides an implementation of the CUDA API that allows multiple processes to use a GPU or many GPUs with better performance than without this mechanism. In case a single process is not able to saturate the whole computing capability of a GPU(s) then many processes can be used with an MPS server as a proxy to the GPU. This allows overlapping computations and data copying from many processes [119] and consequently exploiting a GPU(s) to a higher degree. Additionally, this approach is transparent to the application which is a considerable benefit. MPS requires compute capability

3.5+ and a 64-bit application running under Linux [119]. An MPS daemon can be started as follows:

```
nvidia-cuda-mps-control -d
```

The following experiment tests a scenario without MPS and with MPS for the geometric SPMD application implemented with MPI and CUDA and shown in Section 5.2.4. In every scenario the code was run with various numbers of MPI processes. For every configuration execution time for the best out of three runs is presented in Table 6.1. Application parameters used were `384 384 960 10 2`. Tests were performed on a workstation with 2 x Intel Xeon E5-2620v4, 2 x NVIDIA GTX 1070 GPUs and 128 GB RAM Two GPUs were used.

TABLE 6.1 Execution times [s] for a geometric SPMD MPI+CUDA code, without and with MPS

number of processes	execution time without MPS [s]	execution time with MPS [s]
2	4.668	3.971
4	4.770	3.042
8	7.150	3.339

6.2 DATA GRANULARITY

Data granularity and partitioning may have an an impact on execution time of a parallel application. Specifically, assuming a master-slave application in which input data is divided into data packets which are then distributed among slave processes for computations and then results are gathered, the following would apply:

1. Small data packets would allow good balancing of data among computing nodes/processors. This might be especially useful if there are processors of various computing speed. On the other hand, too many small data packets would result in considerable overhead for communication.

2. Large data packets might result in poor load balancing among computing nodes/processors. Also, in case of really large data packets, computations might start with a delay.

As a result, there is a trade-off and an optimum size of a data packet can be found. This is illustrated in Figure 6.1 for a parallel master-slave numerical integration application for function $f(x) = 10.0/(1.0 + x)$ implemented with MPI and for various numbers of data packets used. Each packet is then partitioned into rectangles of $\frac{1}{40 \cdot 1024}$ width. The test was run on a workstation with 2 x Intel Xeon E5-2620v4 and 128 GB RAM. For each configuration, the best out of 3 runs are shown. Testbed results are shown for 4 and 16 processes of an application.

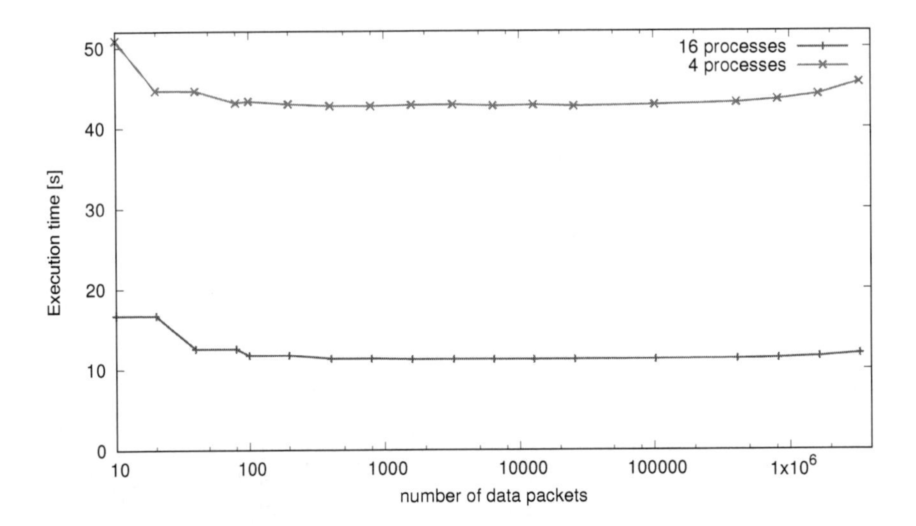

FIGURE 6.1 Execution time of an MPI master-slave application vs number of data packets for fixed input data size

It should be noted that proper distribution of such data among e.g. nodes in a cluster requires efficient load balancing, as discussed in Section 3.1.4.

6.3 MINIMIZATION OF OVERHEADS

6.3.1 Initialization and synchronization overheads

In parallel programming attention should be paid to the costs related to:

1. spawning processes or threads,
2. synchronization.

Specifically, in iterative algorithms such as SPMD implemented with OpenMP, overheads related to thread creation/initialization will be present when entering a `#pragma omp parallel` region. On the other hand, various synchronization constructs discussed in Section 4.2.4 will bring their own delays. In the aforementioned SPMD applications, there are, in particular, two ways of implementation of a loop in which iterations typically correspond to time steps in FDTD simulations [56, 50]. In each loop iteration, the domain is partitioned into subdomains each of which needs to be updated by a separate thread. After updates, synchronization before a following iteration is needed.

Such approaches are also discussed in [23] in the context of OpenMP applications run on an Intel Xeon Phi. In the case of a hybrid MPI+OpenMP application discussed here, a master thread of a process would be involved in communication with other processes. Furthermore, processing of subdomain data assigned to a process must be partitioned into several threads within the process, in this case using OpenMP.

Two functionally identical implementations are as follows:

1. The main loop is executed by the master thread with some parts parallelized using OpenMP constructs. Specifically:
   (a) Exchange of data by the master thread (might be needed first if each process initializes its own domain independently) without any special constructs.
   (b) Parallel update of subdomain cells – parallelization performed using `#pragma omp parallel for`.
   (c) Substitution of pointers for source domain and target domain performed by the master thread without any special constructs.
2. Entering a parallel region outside of the loop. Then each thread would execute loop iterations independently which results in the need for synchronization. Specifically, steps within a loop iteration include:
   (a) Exchange of data by the master thread (might be needed first if each process initializes its own domain independently) in code within `#pragma omp master`.
   (b) Synchronization using `#pragma omp barrier`.
   (c) Parallel update of subdomain cells – parallelization performed using `#pragma omp for`.
   (d) Substitution of pointers for source domain and target domain performed by the master thread in code within `#pragma omp master`.

The two versions of the code can be compiled as follows:

```
mpicc -fopenmp <flags> SPMD-MPI+OpenMP-1.c
mpicc -fopenmp <flags> SPMD-MPI+OpenMP-2.c
```

and executed on a workstation with 2 x Intel Xeon E5-2620v4 and 128 GB RAM with parameters `200 200 200 10000`.

Out of 10 runs for each version, Table 6.2 presents best results for a single process and domain size 200x200x200 and 10000 iterations.

TABLE 6.2 Execution times [s] for two versions of MPI+OpenMP SPMD code

version	minimum execution time [s]	average execution time out of 10 runs [s]
`#pragma omp parallel for` inside main loop	161.362	163.750
`#pragma omp parallel` outside of the main loop	160.349	163.010

6.3.2 Load balancing vs cost of synchronization

In some cases, it might be possible to reduce the time of synchronization at the cost of ability to balance load. For instance, if threads are to fetch input data packets or pointers to input data packets they are supposed to process, the threads would do so in a critical section. In case there are many threads executing, such as for manycore devices such as Intel Xeon Phi, a global critical section might potentially become a bottleneck. A solution to this problem could be to split threads into several groups with the following assumptions and steps:

1. Input data packets are divided into the number of groups equal to the number of thread groups.

2. Instead of one critical section, the number of critical sections equal to the number of thread groups is used, one per group. Then there are fewer threads per critical section which potentially reduces time a thread needs to wait for fetching a new data packet.

In OpenMP, this can be implemented using named critical sections, with a different name for each critical section. A regular implementation would use a regular unnamed critical section.

In terms of results, for runs performed on an Intel Xeon Phi x100 3000 series coprocessor, for a numerical integration application in which each data packet corresponds to a subrange which is further divided into a number of smaller subranges and areas of corresponding rectangles are added, no differences for a version with one critical section as compared to 4 critical sections were observed for up to 32 threads. Then, for 64 threads, the speed-up was better for the latter version by 0.14 up to 58.74 for 128 threads up by 0.8 up to 88.56 and for 228 threads by 1.2 up to 108.78 [45].

6.4 PROCESS/THREAD AFFINITY

In multicore and manycore systems, from performance point of view it might be important to consider how processes or threads of a parallel application are mapped to particular CPUs or CPU cores. It is especially important when threads may share data during computations and proper mapping of threads to cores might efficiently use caches.

In the Linux operating system, it is possible to indicate which CPUs/logical processors in a system will be used for an application. Specifically:

```
taskset -c <cpunumberlist> application
```

will start `application` on the given CPUs in the system. `<cpunumberlist>` contains a comma separated list of CPU numbers starting with `0` e.g. `0,2`, `0-3` etc. In Linux file `/proc/cpuinfo` allows to find out physical ids and core ids of particular logical processors.

In some cases, e.g. when OpenMPI is used, binding can be specified via options to `mpirun`, as shown in Section 4.7.1.

Thread affinity might have a considerable impact on performance results when used on a computing device with a high number of computing cores, such as Intel Xeon Phi [23, 75].

6.5 DATA TYPES AND ACCURACY

Optimization of code also includes the decision on what data types should be used for particular variables in a program. There might be specific functions operating on various types. For instance, the following versions are available for a sine function:

```
float sinf(float a)
double sin(double a)
long double sinl(long double a)
```

Resulting code may offer better performance and, in case of smaller data types, open more potential for vectorization e.g. on Intel Xeon Phi.

Similarly, precision for floating point operations can be controlled with various compiler switches as discussed in [34]. As a result, various trade-offs between execution times and accuracy can be obtained.

In certain applications, smaller data types can be used effectively such as 16-bit fixed point instead of 32-bit floating for training deep networks [79].

6.6 DATA ORGANIZATION AND ARRANGEMENT

It should be noted that proper data arrangement can have an impact on application execution time. If data was prefetched to a cache then it can be accessed much faster.

Let us consider the geometric SPMD example discussed in Section 5.2.1. Specifically, the loops for updates of cells within a subdomain are arranged by indices `z`, `y` and `x`, in that direction. It can be seen from function `getcell` that data cells that are fetched are located next to each other in memory. Let us call this configuration A.

If, for the sake of a test, the order of loops is reversed i.e. indices are browsed in the `x`, `y` and `z` dimension this is no longer the case. Let us call this configuration B.

Table 6.3 presents comparison of execution times of the two configurations for selected numbers of processes of an MPI application, run on a workstation with 2 x Intel Xeon E5-2620v4 and 128 GB RAM. For each configuration, the best out of 3 runs are shown. The codes can be compiled as follows:

```
mpicc <flags> SPMD-MPI-1.c
mpicc <flags> SPMD-MPI-1-xyz-loop.c
```

Tests were run using 16 processes as follows, for a domain of size 600x600x600:

```
mpirun -np 16 ./a.out 600 600 600 100
```

TABLE 6.3 Execution time [s] depending on data layout – 100 iterations of the SPMD code

domain size X×Y×Z	data layout A	data layout B
600×600×600	23.81	95.081
800×800×800	55.236	236.032
1000×1000×1000	106.808	460.754

Often, tiling or blocking data [110] allows reusing of data that has been loaded into cache or registers. For instance, if a loop stores data in a large array (that does not fit into cache) such that in each iteration a successive element is stored and then this data is used for subsequent updates of other data, such a large loop can be tiled into a few passes each of which reuses data from the cache.

6.7 CHECKPOINTING

Checkpointing is a mechanism that allows saving of the state of an application and resume processing at a later time. Saving the state of a parallel application is not easy because it requires saving a global consistent state of many processes and/or many threads, possibly running on various computing devices, either within a node or on a cluster, with consideration of communication and synchronization.

Checkpointing might be useful for maintenance of the hardware. In some cases, it may also allow moving the state of an application or individual application processes to other locations. This effectively implements migration. The latter might be useful for minimization of application execution times when processes perform computations for which execution times cannot be predicted in advance such as in alpha beta search.

In general, checkpointing can be implemented at various levels [167], in particular provided by:

1. Hardware.
2. Operating system kernel.
3. User-level [49] in which an existing library is linked with an executable to provide checkpointing.
4. Application level checkpointing [167] which requires a programmer's effort but can result in higher performance as only data really necessary to carry on execution need to be stored in an application state [49].

Many works concerning checkpointing on modern high performance computing systems have appeared, suitable for applications using particular, previously discussed parallel programming APIs:

- CPPC (ComPiler for Portable Checkpointing) [140] allows checkpointing of parallel message passing applications with independence from the operating system and the message passing protocol. Code transformation is done transparently by a provided compiler. The solution works at the level which requires only saving variables necessary for restart. The compiler inserts checkpoints at identified safe points.
- Paper [107] presents extension of CPPC for checkpointing of hybrid MPI+OpenMP applications with a protocol that applies coordinated checkpointing among threads of a team and proper analysis of communication.
- CheCUDA [154] allows checkpointing of CUDA applications using an approach with transferring data from device to host memory, checkpointing, reinitializing and transferring data back to the device. Berkeley Lab Checkpoint/Restart (BLCR) [81] is used for checkpoint and restart on the host side.
- NVCR [117] – a solution for checkpointing CUDA programs that is more transparent than CheCUDA as the former replaces CUDA libraries with proper wrappers. NVCR allows storing of data associated with CUDA and restoring it after restart. It can work with MPI+CUDA applications.
- CheCL [153] – a solution for checkpointing OpenCL applications that replaces an original OpenCL library which allows intercepting original calls. CheCL uses an API proxy process to which original calls are forwarded and which makes actual OpenCL calls. This allows storing needed data in the original process and checkpointing it. BLCR is used. The approach can be used with MPI applications. It was shown that the approach allows migration of a process from one node to another.

- Application level checkpointing for OpenMP applications [26]. A programmer is required to denote checkpoint points in the code. The C^3 precompiler is used to for code instrumentation and inserting code for checkpointing. According to the authors this solution can be combined with their approach to checkpointing MPI applications [25]. The latter is also based on a instrumenting the source code with the precompiler for subsequent checkpointing at the application level. A coordination layer is used to intercept MPI invocations from the application and implement non-blocking, coordinated protocol to achieve global checkpointing.
- Hybrid Kernel Checkpoint that can save and restore a GPU kernel state [151].
- Checkpointing using NVRAM for MPI applications – paper [62] proposes how to use new technologies such as NVRAMs located in cluster nodes in order to optimize application execution with checkpointing.

6.8 SIMULATION OF PARALLEL APPLICATION EXECUTION

As the sizes of modern HPC systems have increased considerably and there is a great variety of configurations that can be built with many components such as CPUs, GPUs, coprocessors and network interconnects, it is not straightforward to choose the best hardware platform for a given application or a set of applications. Furthermore, in some cases it would be beneficial to test an application for a larger size (such as in terms of the number of nodes, CPUs, cores) compared to an available configuration or potential configurations considered for purchase – such as in case of a considered upgrade. In such scenarios, it is possible to use one of the simulation environments that allow simulation of the execution of an application (can be modeled with a dedicated language or derived from existing applications) on a system that consists of a number of computing devices connected with a particular network. In general, there can be several use cases in which such an environment can be helpful, in particular [52]:

1. Prediction of application performance on a larger system e.g. to assess potential benefits.
2. Identification of bottlenecks and tuning of an application by testing various configurations including sizes of buffers, algorithm improvements etc.
3. Assessment of the best system to run the actual application.

Examples of such systems and use cases applicable in the context of this book include:

- MERPSYS [52] – for modelling and simulation of execution time, energy consumption and reliability of parallel applications run on cluster and volunteer based systems.
- MARS [58] – for performance prediction and tuning of MPI applications running on systems with various network topologies.
- SST/macro [8] – for performance analysis of parallel applications run on a parallel system with computing and interconnect components.

6.9 BEST PRACTICES AND TYPICAL OPTIMIZATIONS

6.9.1 GPUs/CUDA

Typically, several techniques can be used to ensure that high performance of computations can be achieved on a GPU, some of which include:

1. Dynamic scheduling of warps on available multiprocessors of a GPU – handled by the runtime layer.
2. Minimization of thread divergence – making sure that possibly all threads in a warp take the same execution path.
3. Caching that can be done through using shared memory – data is copied from global to shared memory first, then computations are applied and results copied back to global memory and RAM.
4. Prefetching data/hiding communication latency – this technique tries to avoid idle times between the moment previous computations completed and following computations for which new input data needs to be copied. In essence, new input data is prefetched while previous computations are still running. In the context of GPU processing, this technique applies to the following:

   (a) Fetching input data from global memory which is reasonably slow compared to shared memory and registers. If an algorithm works on data chunks in a loop, data can be first loaded from global memory to registers, and then in a loop the following actions can be performed: a data chunk is loaded from registers to shared memory, synchronization among threads within a block is performed, a new data chunk is fetched from global memory to registers, computations are performed on the current data chunk, synchronization among threads within a block is performed and a new loop iteration is executed. This approach allows overlapping computations on a current data chunk with prefetching of a new one.

(b) communication between the host and the GPU(s). If multiple streams are used and capable GPUs are used, it is possible to overlap communication between the host and the device (and possibly between the device and the host) with kernel execution on the device. This is discussed in Section 6.1.2.

5. Memory coalescing – from the performance point of view, it is best if threads executing in parallel read/write data from successive memory location of the global memory; in some cases proper reorganization of data and/or code can consequently result in shorter execution time.

6. Accessing various shared memory banks from warp threads. Shared memory is divided into banks, each of which can be accessed in parallel. In general, if requests from threads are directed to one bank, accesses will be serialized which affects performance. An n-way bank conflict occurs if n threads access the same bank at the same time. Documentation [122] describes how shared memory organization for particular compute capabilities which will have an impact on performance. For instance, for compute capability 5.x devices, there are 32 banks with mapping of successive 32-bit words to various banks. If two threads of a warp access data within one word, no conflict occurs either for read or write operations [122].

7. Loop unrolling.

8. Minimization of thread synchronization in the algorithm.

6.9.2 Intel Xeon Phi

The architecture of the Intel Xeon Phi manycore solution requires from a programmer to pay attention to several issues that have considerable impact on potential speed-up and performance [135, 110, 31, 5, 23, 75, 166] of an application running on such a system. General guidelines would be to investigate the following checklist – make sure that:

1. Code is highly parallel i.e. it exposes enough parallelism to make use of the several tens+ physical cores of the processor and the possibility to run a few hardware threads per core. In fact, at least two threads (sometimes 3 or 4) per core should be engaged in order to make use of the full potential of the coprocessor [135]. In practice, this requires that for a considerable percentage of the running time of the application, a large number of threads should be running in parallel.

2. Make sure that vectorization is used within the code as much as possible. This can be relied on thanks to automatic parallelization by a compiler. Intel's `icc` compiler allows compiling applications targeted for Xeon Phi

with, in particular, support for vectorization. The `icc` compiler will attempt vectorization and comments on its output can be checked when providing compilation flag `-vec-report<n>` where `n` is equal or greater than 1, the higher the `n`, the more information is displayed. Section 4.2.7 presents OpenMP directives that can be used for declaration of execution using SIMD instructions. There are several pragmas accepted by the Intel `icc` compiler that allow the indication to ignore dependencies and enforcing vectorization [110].

3. Use alignment of data structures in memory.

4. Cache is utilized efficiently among threads running within a core and between cores.

5. Proper thread affinity is used. Article [76] discusses how threads can be mapped to cores with either OpenMP affinity environment variables (discussed in Section 4.2.5) or Intel environment variables. The latter includes `KMP_AFFINITY` that specifies assignment of threads in OpenMP to hardware threads with granularity: fine – in which case each thread is bound to a single hardware thread, core – in which case four threads form a group which is bound to a core as well as affinity which can be compact, scatter and also balanced (the latter specific to Intel Xeon Phi):

   - compact – the next thread is assigned beside a previous one considering hardware threads within cores of a Xeon Phi,
   - scatter – the next thread will be assigned to the next physical core,
   - balanced – all threads are divided into possibly equal sized groups (with threads in a group with successive ids) and assigned to physical cores of the Intel Xeon Phi.

   `KMP_PLACE_THREADS` allows setting affinity more precisely using the following fields: `C` – cores, `T` – threads, `O` – denotes an offset in cores starting from core 0. For example, `export KMP_PLACE_THREADS=32C,3T,16O` will schedule 3 threads per core on a total of 32 cores starting from core 16. The default value for the offset is `0O`.

6. False sharing is minimized. Each core of the system has a cache memory. As cache needs to be coherent among the cores, it may result in an overhead. For instance, if threads executing on various cores update array cells that are in one cache line, this will result in overhead due to cache lines that need to be coherent among nodes. Consequently, if various threads modify data in memory locations not even overlapping but in close proximity, falling into one cache line, this may result in considerable overhead across the system. So a code, while formally correct, may not execute as fast as it could if it was organized differently. As an example, in case of array elements of which are updated by several

threads, there is a risk of false sharing. There are a few ways of dealing with false sharing, including:

(a) Instead of updates in locations that may result in false sharing, each thread could use its own local copies with private variables and only update final memory space at the end of computations. This may potentially increase memory usage but may result in shorter execution time.

(b) Padding a type of an element updated by a thread so that the whole type is a multiple of the size of a cache line. Similarly to the first solution, this leads to shorter execution time at the cost of higher memory requirements.

7. Memory per thread is used efficiently.
8. Synchronization among threads is kept to minimum and optimum synchronization methods are used. For instance, refer to Section 6.3.2.

In case of Intel Xeon Phi x200, efficient execution of parallel codes will also involve decisions on the following (article [74] discusses these modes in more detail along with potential use cases):

1. memory mode that defines which type of memory (DRAM, MCDRAM) and how they are visible to an application:
    - Flat – both DRAM and MCDRAM are available for memory allocation (as NUMA nodes), the latter preferably for bandwidth critical data.
    - Cache – in this case MCDRAM acts as an L3 cache.
    - Hybrid – a part of MCDRAM is configured as cache and a part as memory that can be allocated by an application.
    - MCDRAM – only MCDRAM is available.
2. cluster mode that defines how requests to memory are served through memory controllers:
    - All2All (default in case of irregular DIMM configuration) – a core, a tag directory (to which a memory request is routed) and a memory channel (to which a request is sent in case data is not in cache) can be in various parts.
    - Quadrant (in case of symmetric DIMM configuration) – a tag directory and a memory channel are located in the same region.
    - Sub-NUMA clustering – in this mode regions (quarter – SNC4, half – SNC2) will be visible as separate NUMA nodes. A core, a tag directory and a memory channel are located in the same region

which potentially result in optimization if handled properly through affinity settings such as assignment of various processes to various nodes.

Apart from OpenMP constructs discussed in Section 4.2.8, offloading computations to an Intel Xeon Phi coprocessor is possible with the `offload` directive recognized by the Intel compiler, with the possibility of specifying space allocation and release on the device, data copying between the host and the device, specifying a particular device (in case several Xeon Phi cards are used) and also asynchronous launching of computations on a device to overlap the latter with computations on the host [57].

Regarding fabrics for communication used by MPI, settings for both intranode and internode can be selected using the `I_MPI_FABRICS` environment variable [94, Selecting Fabrics] such as `export I_MPI_FABRICS=<intranode fabric>:<internode fabric>`.

Work [24] discusses performance of various data representations, namely Structure of Arrays (SoA) and AoS (Array of Structures) when running parallel vectorized code on Intel Xeon and Intel Xeon Phi. As a conclusion, in terms of performance generally SoA should be preferred. Generally, it produces fewer instructions.

6.9.3 Clusters

For clusters, that consist of many nodes, the following techniques should be employed during development of parallel programs:

1. Expose enough parallelism in the application considering the number of computing devices: CPUs, streaming multiprocessors within GPUs or coprocessors.
2. Design parallelization and data granularity in a data partitioning algorithm in a way that takes into account performance differences between computing devices. Specifically, small data packets distributed among computing devices allow for fine grained load balancing but such a strategy involves more overheads for handling packets and results. On the other hand, large data packets do not allow for good load balancing in a heterogeneous environment and result in larger execution times [39, 141].
3. Optimize communication taking into account:
   - overlapping computations and communication,
   - minimization of communication operations and piggybacking i.e. in some cases information related to dynamic load balancing can be added to the data of the algorithm itself.

4. Optimize cache usage such that as much data as possible in consecutive computations is used from a nearby cache.

5. Assess whether idle times can be hidden in the algorithm. Specifically, for some so-called irregular problems, some processes or threads may run out of data or computations unexpectedly which leads to inefficient use of computing devices. It might be possible to hide those idle times e.g. by spawning two or more processes per core to minimize this risk at the cost of some overhead.

6. Design and implement a load balancing strategy. Note that there can be a trade-off between the cost of load balancing and potential gains. Specifically, a load balancing algorithm may try to maintain approximately the same load (within certain ranges) on all nodes with high frequency which would minimize the risk of idle times showing up on the nodes. However, this process also consumes CPU cycles and entails communication costs for exchanging load information. On the other hand, waiting too long with load balancing does not involve these overheads but may result in idle times until computing devices become active again.

7. Implement checkpointing for long running applications, as described in Section 6.7.

6.9.4 Hybrid systems

Hybrid systems can be thought of as systems with different processors that either require programming using various APIs and/or are located at various levels in terms of system architecture. Some processors might be better suited for specific types of codes and consequently might require specific optimizations and differ in performance and often power requirements. An example would be a multicore CPU and a GPU, a multicore CPU and a Xeon Phi hybrid system. It should also be noted that integration of several nodes of such types into a cluster creates a multilevel, hybrid and heterogeneous system in which parallelization must be implemented among nodes and within nodes.

Apart from specific optimizations valid for particular types of systems, a hybrid, heterogeneous system requires paying attention and optimization of the following:

1. Data granularity, data partitioning and load balancing, minimization of synchronization overheads. Often there are optimal batch sizes (Section 6.2) that can even be adjusted at runtime for lowest execution time [46].

2. Management of computations at lower levels. Specifically, within a node it is preferable to dedicate CPU cores to threads that will manage computations on computing devices such as GPUs or Xeon Phi cards [46].

3. Overlapping communication and computations, such as:

- overlapping computations on CPUs and internode communication using non-blocking MPI calls or threads,
- overlapping computations on accelerators such as GPU and host-device communication e.g. using streams with CUDA (Section 6.1.2).

4. Potentially various affinities can be used for various devices. Specifically `MIC_ENV_PREFIX` allows setting a prefix and then an affinity for a Xeon Phi in such a case, as follows:

```
export MIC_ENV_PREFIX=PHI
export PHI_KMP_AFFINITY=balanced
export PHI_KMP_PLACE_THREADS=60c,3t
export PHI_OMP_NUM_THREADS=180
```

5. Multilevel parallel programming can be naturally achieved with a combination of APIs discussed in this book e.g. MPI for internode communication and e.g. OpenMP for management of computations within a node with spawning work on various devices such as GPUs with CUDA, and nested parallelism for parallelization among CPU cores.

Performance wise, best results in a hybrid environment are achieved for various devices that offer reasonably similar performance. Otherwise, a parallel implementation might struggle to get a considerable performance gain from adding slower devices to a system.

APPENDIX A

Resources

CONTENTS

A.1 SOFTWARE PACKAGES

This appendix describes installation of software needed for compilation and execution of parallel applications contained in this book. The following software packages were used for this purpose under a Linux operating system (Ubuntu distribution):

- gcc installed from packages using a package manager,
- NVIDIA CUDA environment – installed from packages using a package manager:

  ```
  sudo apt install nvidia-cuda-toolkit
  ```

- OpenMPI installed as follows, after an OpenMPI archive has been downloaded:

  ```
  tar xvzf openmpi-<version>.tar.gz
  cd openmpi-<version>/
  ./configure --prefix=/home/pczarnul/openmpi-<version>-bin \
  --enable-mpi-thread-multiple --with-cuda=/usr
  make all install
  ```

  as well as exporting a path to MPI binaries:

  ```
  export PATH=/home/pczarnul/openmpi-<version>-bin/bin:$PATH
  ```

  in file ~/.bashrc

- OpenACC – accULL version 0.4 alpha software was used for compilation and running of the testbed example.

APPENDIX B

Further reading

CONTENTS

B.1 CONTEXT OF THIS BOOK

This book combines descriptions of key APIs (with the most focus on MPI, OpenMP, CUDA and OpenCL) used for high performance computing and outlines key elements from the point of view of the contribution in this book i.e.:

1. implementation of commonly found paradigms such as master-slave, geometric SPMD and divide-and-conquer using these APIs,
2. hybrid implementations using combinations of selected APIs suitable for efficient use of modern hybrid HPC systems.

Additionally, selected useful modern elements of these APIs were presented such as: dynamic parallelism and unified memory in CUDA, dynamic process creation, parallel I/O, one-sided API in MPI, offloading, tasking in OpenMP.

B.2 OTHER RESOURCES ON PARALLEL PROGRAMMING

There are several sources that can be explored for further reading. These include:

1. Tutorials and presentations on particular APIs such as for:
   - OpenMP: [22] that includes, in particular, discussion of OpenMP support by various compilers as well as discussion of C and Fortran APIs.
   - MPI: [21] with discussion of various MPI implementations, C and Fortran APIs and exercises support. Presentation [36] includes discussion of communicators (also intercommunicators), groups, collective communication across intercommunicators, one-sided communication, parallel I/O, performance analysis with profiling using

the TAU Performance System® [59, 150] and Vampir/Vampirtrace. Currently Vampir can be obtained at [80].

- Pthreads: [7] that includes, in particular, comparison of execution times of functions `fork()` and `pthread_create()` on various platforms as well as comparison of memory bandwidths for MPI and Pthreads. Additionally, apart from the discussion of the API, the tutorial includes a section on monitoring and debugging Pthreads applications.
- OpenCL: [95] with links to several tutorials related to OpenCL. [11] provides description of the AMD Accelerated Parallel Processing Implementation of OpenCL.
- OpenACC: [124] provides description of programming APIs, examples as well as best practices. In particular, the following are covered: evaluation of application performance, parallelization of loops, data locality considerations, loop optimization, programming using many devices as well as asynchronous operations.

2. Books such as on:

- CUDA: [33] that includes discussion of programming API, GPU accelerated libraries, debugging and profiling.
- CUDA: [101] with many examples, but also MPI and OpenACC.
- OpenACC: [66] with profile driven application development, data management discussion, many examples of applications using OpenACC.
- Optimization of many real life problems: [136] including, among others, a hydro2D, solving Navier-Stokes equations, deep learning, N-body problem, 3D finite differences algorithm etc. Selected APIs and computing devices are discussed. Book [97] discusses more examples and optimization of, among others, the following: numerical weather prediction, pairwise DNA sequence alignment, numerical methods, visual search, matrix computations etc. In particular, the book discusses hStreams and shared memory programming in MPI 3, OpenCL, performance considerations in OpenMP vs OpenCL, SIMD functions in OpenMP, vectorization etc.
- Shared memory programming: [12] with discussion of, among others, cache coherency and consistency of memory. The book considers Pthreads, Windows® threads, C++11 Thread Library, OpenMP, Intel Threading Building Blocks.
- Multicore and GPU programming: [20] with consideration of selected computing hardware including, among others, NVIDIA GPUs, AMD APUs, multi and many core, patterns including, in particular: SPMD, MPMD, master-worker, map-reduce and

fork-join, as well as APIs including, in particular: OpenMP, MPI (master-worker considered as a case study), CUDA, Thrust, Boost.MPI library.

- Multicore software development: [130] including, among others, thread API discussion with, in particular, C++, Java, OpenMP, .NET, Intel Threading Building Blocks, Microsoft® Task Parallel Library, PLINQ, CUDA. The book discusses emerging solutions including, in particular, automatic parallelism extraction, transactional memory, auto tuning and future applications.

3. Teaching high performance computing and environments that support such activities. There are several works that address content, structure, approaches, experiences and suggestions for the future in terms of relevant courses e.g. [87, 149, 144]. Several works describe software that can be used for teaching parallel programming, for instance:

   - OnRamp [68] which is a web portal that allows teaching concepts in parallel and distributed computing, getting acquainted with parallel code, parallel computing environment and accessing the machine. This is possible through knowledge levels that introduce elements level by level. It is possible to assign codes in MPI, CUDA, OpenMP to a Module with various parameters and configuration.
   - BeesyCluster [42] is a middleware that allows access to distributed servers, workstations and clusters including exposing serial/parallel codes as services which can then be integrated internally into workflow applications with various scheduling algorithms. BeesyCluster, thanks to its features, has also been used for teaching high performance computing systems programming. It allows WWW and Web Service based access to clusters through a BeesyCluster account to which many user accounts can be linked. The WWW interface allows creation of files and directories, editing files, compilation, running interactively or using a queueing system, working in a group with facilities such as an internal messaging system and a shared board.
   - A grid portal for teaching parallel programming courses, with its multi-layer architecture and experiences during design and implementation was presented in [156]. It also allows access to heterogeneous resources located in various institutions and was used for an undergraduate course.

Bibliography

[1] Comcute Website. Department of Computer Architecture, Faculty of Electronics, Telecommunications and Informatics, Gdansk University of Technology, http://comcute.eti.pg.gda.pl/, accessed on 19th July 2017.

[2] GCC Wiki: OpenACC. https://gcc.gnu.org/wiki/OpenACC, accessed on 19th July 2017.

[3] Hadoop Wiki. CUDA On Hadoop. https://wiki.apache.org/hadoop/CUDA%20On%20Hadoop, accessed on 19th July 2017.

[4] TOP500 Supercomputer Sites. Editors: H. Meuer, E. Strohmaier, J. Dongarra, H. Simon, M. Meuer, http://top500.org/, last accessed on 19th July 2017.

[5] Avoiding and Identifying False Sharing among Threads, November 2011. Intel Developer Zone, https://software.intel.com/en-us/articles/avoiding-and-identifying-false-sharing-among-threads, accessed on 19th July 2017.

[6] MPI : A Message-Passing Interface Standard, Version 3.1, June 2015. Message Passing Interface Forum, http://www.mpi-forum.org/docs/mpi-3.1/mpi31-report.pdf, accessed on 19th July 2017.

[7] POSIX Threads Programming, March 2017. Lawrence Livermore National Laboratory, https://computing.llnl.gov/tutorials/pthreads/, accessed on 19th July 2017.

[8] H. Adalsteinsson, S. Cranford, D. A. Evensky, J. P. Kenny, J. Mayo, A. Pinar, and C. L. Janssen. A Simulator for Large-Scale Parallel Computer Architectures. *Int. J. Distrib. Syst. Technol.*, 1(2):57–73, Apr. 2010.

[9] A. Adinetz. Adaptive Parallel Computation with CUDA Dynamic Parallelism, May 2014. https://devblogs.nvidia.com/parallelforall/introduction-cuda-dynamic-parallelism/, accessed on 19th July 2017.

[10] A. Adinetz. CUDA Dynamic Parallelism API and Principles, May 2014. `https://devblogs.nvidia.com/parallelforall/cuda-dynamic-parallelism-api-principles/`, accessed on 19th July 2017.

[11] Advanced Micro Devices. AMD Accelerated Parallel Processing OpenCL Programming Guide, November 2013. `http://developer.amd.com/wordpress/media/2013/07/AMD_Accelerated_Parallel_Processing_OpenCL_Programming_Guide-rev-2.7.pdf`, accessed on 19th July 2017.

[12] V. Alessandrini. *Shared Memory Application Programming: Concepts and Strategies in Multicore Application Programming.* Morgan Kaufmann, Waltham, MA, USA, November 2015. ISBN 978-0128037614.

[13] D. Anderson. BOINC: A System for Public-Resource Computing and Storage. In *Proceedings of 5th IEEE/ACM International Workshop on Grid Computing*, Pittsburgh, USA, November 2004.

[14] D. P. Anderson, E. Korpela, and R. Walton. High-Performance Task Distribution for Volunteer Computing. In *Proceedings of the First International Conference on e-Science and Grid Computing*, E-SCIENCE '05, pages 196–203, Washington, DC, USA, 2005. IEEE Computer Society.

[15] P. Balaji and T. Hoefler. Advanced Parallel Programming with MPI-1, MPI-2, and MPI-3, February 2013. PPoPP, Shenzhen, China, `https://htor.inf.ethz.ch/teaching/mpi_tutorials/ppopp13/2013-02-24-ppopp-mpi-advanced.pdf`, accessed on 19th July 2017.

[16] J. Balicki, H. Krawczyk, and E. Nawarecki, editors. *Grid and Volunteer Computing.* Gdansk University of Technology, Faculty of Electronics, Telecommunication and Informatics Press, Gdansk, 2012. ISBN: 978-83-60779-17-0.

[17] R. Banger and K. Bhattacharyya. *OpenCL Programming by Example.* Packt Publishing, December 2013. ISBN: 978-1849692342.

[18] A. Barak, T. Ben-nun, E. Levy, and A. Shiloh. A Package for OpenCL Based Heterogeneous Computing on Clusters with Many GPU Devices. In *Proc. of Int. Conf. on Cluster Computing*, pages 1–7, September 2011.

[19] A. Barak and A. Shiloh. The VirtualCL (VCL) Cluster Platform, 2017. white paper, `http://www.mosix.cs.huji.ac.il/vcl/VCL_wp.pdf`, accessed on 26th July 2017.

[20] G. Barlas. *Multicore and GPU Programming: An Integrated Approach.* Morgan Kaufmann, Waltham, MA, USA, 1st edition, December 2014. ISBN 978-0124171374.

[21] B. Barney. Message Passing Interface (MPI), June 2017. Lawrence Livermore National Laboratory, https://computing.llnl.gov/tutorials/mpi/, accessed on 19th July 2017.

[22] B. Barney. OpenMP, June 2017. Lawrence Livermore National Laboratory, https://computing.llnl.gov/tutorials/openMP/, accessed on 19th July 2017.

[23] M. Barth, M. Byckling, N. Ilieva, S. Saarinen, M. Schliephake, and V. Weinberg. Best Practice Guide Intel Xeon Phi, February 2014. v 1.1, Partnership for Advanced Computing in Europe, http://www.prace-ri.eu/best-practice-guide-intel-xeon-phi-html/, accessed on 19th July 2017.

[24] P. Besl. A Case Study Comparing AoS (Arrays of Structures) and SoA (Structures of Arrays) Data Layouts for a Compute-intensive Loop Run on Intel Xeon Processors and Intel Xeon Phi Product Family Coprocessors. https://software.intel.com/sites/default/files/article/392271/aos-to-soa-optimizations-using-iterative-closest-point-mini-app.pdf, accessed on 19th July 2017.

[25] G. Bronevetsky, D. Marques, K. Pingali, and P. Stodghill. Automated Application-level Checkpointing of MPI Programs. In *Proceedings of the Ninth ACM SIGPLAN Symposium on Principles and Practice of Parallel Programming*, PPoPP '03, pages 84–94, New York, NY, USA, 2003. ACM.

[26] G. Bronevetsky, K. Pingali, and P. Stodghill. Experimental Evaluation of Application-level Checkpointing for OpenMP Programs. In *Proceedings of the 20th Annual International Conference on Supercomputing*, ICS '06, pages 2–13, New York, NY, USA, 2006. ACM.

[27] S. Buckeridge and R. Scheichl. Parallel Geometric Multigrid for Global Weather Prediction. *Numerical Linear Algebra with Applications*, 17(2-3):325–342, 2010.

[28] J. Burkardt. FD2D HEAT EXPLICIT SPMD Finite Difference 2D Heat Equation Using SPMD, April 2010. https://people.sc.fsu.edu/~jburkardt/m_src/fd2d_heat_explicit_spmd/fd2d_heat_explicit_spmd.html, accessed on 19th July 2017.

[29] R. Buyya. *High Performance Cluster Computing: Programming and Applications*. Prentice Hall PTR, Upper Saddle River, NJ, USA, 1st edition, 1999.

[30] R. Buyya, S. Venugopal, R. Ranjan, and C. S. Yeo. *Market Oriented Grid and Utility Computing*, chapter The Gridbus Middleware for Market-Oriented Computing. Wiley Press, Hoboken, New Jersey, USA, October 2009. ISBN: 978-0470287682.

[31] S. Cepeda. Optimization and Performance Tuning for Intel Coprocessors, Part 2: Understanding and Using Hardware Events, November 2012. Intel Developer Zone, https://software.intel.com/en-us/articles/optimization-and-performance-tuning-for-intel-xeon-phi-coprocessors-part-2-understanding, accessed on 19th July 2017.

[32] B. Chapman, G. Jost, and R. v. d. Pas. *Using OpenMP: Portable Shared Memory Parallel Programming (Scientific and Engineering Computation).* The MIT Press, Cambridge, MA, 2007.

[33] J. Cheng, M. Grossman, and T. McKercher. *Professional CUDA C Programming.* Wrox, 1st edition, September 2014. ISBN 978-1118739327.

[34] D. M. J. Corden and D. Kreitzer. Consistency of Floating-Point Results using the Intel Compiler or Why Doesnt My Application Always Give the Same Answer?, September 2017. Software Solutions Group, Intel Corporation, `https://software.intel.com/en-us/articles/consistency-of-floating-point-results-using-the-intel-compiler`.

[35] T. H. Cormen, C. E. Leiserson, R. L. Rivest, and C. Stein. *Introduction to Algorithms, Third Edition.* The MIT Press, Cambridge, MA, 3rd edition, 2009.

[36] D. Cronk. Advanced MPI, 2004. Innovative Computing Lab, University of Tennessee, `https://hpc.llnl.gov/sites/default/files/DavidCronkSlides.pdf`, accessed on 19th July 2017.

[37] R. Cushing, G. Putra, S. Koulouzis, A. Belloum, M. Bubak, and C. de Laat. Distributed Computing on an Ensemble of Browsers. *Internet Computing, IEEE*, 17(5):54–61, 2013.

[38] P. Czarnul. Programming, tuning and automatic parallelization of irregular divide-and-conquer applications in DAMPVM/DAC. *International Journal of High Performance Computing Applications*, 17(1):77–93, Spring 2003.

[39] P. Czarnul. Parallelization of Compute Intensive Applications into Workflows based on Services in BeesyCluster. *Scalable Computing: Practice and Experience*, 12(2):227–238, 2011.

[40] P. Czarnul. A Model, Design and Implementation of an Efficient Multithreaded Workflow Execution Engine with Data Streaming, Caching, and Storage Constraints. *Journal of Supercomputing*, 63(3):919–945, MAR 2013.

[41] P. Czarnul. Modeling, Run-Time Optimization and Execution of Distributed Workflow Applications in the Jee-Based Beesycluster Environment. *Journal of Supercomputing*, 63(1):46–71, JAN 2013.

[42] P. Czarnul. Teaching High Performance Computing Using BeesyCluster and Relevant Usage Statistics. *Procedia Computer Science*, 29:1458 – 1467, 2014. 2014 International Conference on Computational Science.

[43] P. Czarnul. BeesyCluster as Front-End for High Performance Computing Services. *TASK Quarterly, Scientific Bulletin of the Academic Computer Centre in Gdansk*, 19(4):387396, 2015. ISSN 1428-6394, `http://task.gda.pl/files/quart/TQ2015/04/tq419f-c.pdf`.

[44] P. Czarnul. *Integration of Services into Workflow Applications*. Chapman and Hall/CRC, Boca Raton, FL, USA, 2015.

[45] P. Czarnul. *Parallel programming on Intel Xeon Phi. Compute intensive parallel applications using OpenMP and MPI.* Gdansk University of Technology, 2015.

[46] P. Czarnul. Benchmarking Performance of a Hybrid Intel Xeon/Xeon Phi System for Parallel Computation of Similarity Measures Between Large Vectors. *International Journal of Parallel Programming*, pages 1–17, 2016.

[47] P. Czarnul. *Information Systems Architecture and Technology: Proceedings of 36th International Conference on Information Systems Architecture and Technology – ISAT 2015 – Part III*, chapter Parallelization of Divide-and-Conquer Applications on Intel Xeon Phi with an OpenMP Based Framework, pages 99–111. Springer International Publishing, Cham, Switzerland, 2016.

[48] P. Czarnul, M. Bajor, M. Fraczak, A. Banaszczyk, M. Fiszer, and K. Ramczykowska. Remote Task Submission and Publishing in BeesyCluster: Security and Efficiency of Web Service Interface. In Wyrzykowski, R. and Dongarra, J. and Meye, N. and Wasniewski, J., editor, *Parallel Processing and Applied Mathematics*, volume 3911 of *Lecture Notes in Computer Science*, pages 220–227, 2006. 6th International Conference on Parallel Processing and Applied Mathematics, Poznan, Poland, Sep 11-14, 2005.

[49] P. Czarnul and M. Fraczak. New User-Guided and Ckpt-Based Checkpointing Libraries for Parallel MPI Applications. In DiMartino, B and Kranzlmuller, D and Dongarra, J, editor, *Recent Advances in Parallel Virtual Machine and Message Passing Interface, Proceedings*, volume 3666 of *Lecture Notes in Computer Science*, pages 351–358, 2005. 12th European Parallel-Virtual-Machine-and-Message-Passing-Interface-Users-Group Meeting (PVM/MPI), Sorrento, Italy, Sep 18-21, 2005.

[50] P. Czarnul and K. Grzeda. Parallel Simulations of Electrophysiological Phenomena in Myocardium on Large 32 and 64-bit Linux Clusters.

In Kranzlmuller, D. and Kacsuk, P. and Dongarra, J., editor, *Recent Advances In Parallel Virtual Machine And Message Passing Interface, Proceedings*, volume 3241 of *Lecture Notes In Computer Science*, pages 234–241, 2004. 11th European Parallel Virtural Machine and Message Passing Interface Users Group Meeting, Budapest, Hungary, Sep 19-22, 2004.

[51] P. Czarnul, J. Kuchta, and M. Matuszek. *Parallel Computations in the Volunteer–Based Comcute System*, pages 261–271. Springer Berlin Heidelberg, Berlin, Heidelberg, 2014.

[52] P. Czarnul, J. Kuchta, M. Matuszek, J. Proficz, P. Rosciszewski, M. Wojcik, and J. Szymanski. MERPSYS: An Environment for Simulation of Parallel Application Execution on Large Scale HPC Systems. *Simulation Modelling Practice and Theory*, 77:124 – 140, 2017.

[53] P. Czarnul and M. Matuszek. Performance Modeling and Prediction of Real Application Workload in a Volunteer-based System. In *Applications of Information Systems in Engineering and Bioscience, Proceedings of 13th International Conference on Software Engineering, Parallel and Distributed Systems conference (SEPADS)*, pages 37–45, Gdansk, Poland, May 2014. WSEAS. ISBN: 978-960-474-381-0, `http://www.wseas.us/e-library/conferences/2014/Gdansk/SEBIO/SEBIO-03.pdf`.

[54] P. Czarnul and M. Matuszek. Considerations of Computational Efficiency in Volunteer and Cluster Computing. In *Proceedings of Parallel Processing and Applied Mathematics (PPAM)*, Lecture Notes in Computer Science. Springer, 2015.

[55] P. Czarnul and P. Rosciszewski. Optimization of Execution Time under Power Consumption Constraints in a Heterogeneous Parallel System with GPUs and CPUs. In M. Chatterjee, J.-n. Cao, K. Kothapalli, and S. Rajsbaum, editors, *Distributed Computing and Networking*, volume 8314 of *Lecture Notes in Computer Science*, pages 66–80. Springer Berlin Heidelberg, 2014.

[56] P. Czarnul, S. Venkatasubramanian, C. Sarris, S. Hung, D. Chun, K. Tomko, E. Davidson, L. Katehi, and B. Perlman. Locality Enhancement and Parallelization of an FDTD Simulation. In *Proceedings of the 2001 DoD/HPCMO Users Conference*, U.S.A., 2001.

[57] K. Davis. Effective Use of the Intel Compiler's Offload Features, September 2013. `https://software.intel.com/en-us/articles/effective-use-of-the-intel-compilers-offload-features`, accessed on 19th July 2017.

[58] W. E. Denzel, J. Li, P. Walker, and Y. Jin. A Framework for End-to-end Simulation of High-performance Computing Systems. In *Proceedings of the 1st International Conference on Simulation Tools and Techniques for Communications, Networks and Systems & Workshops*, Simutools '08, pages 21:1–21:10, ICST, Brussels, Belgium, Belgium, 2008. ICST (Institute for Computer Sciences, Social-Informatics and Telecommunications Engineering).

[59] Department of Computer and Information Science, University of Oregon, Advanced Computing Laboratory, LANL, NM, Research Centre Julich, ZAM, Germany. TAU Performance System, June 2017. `http://www.cs.uoregon.edu/research/tau/home.php`, accessed on 19th July 2017.

[60] J. Dongarra. Emerging Heterogeneous Technologies for High Performance Computing, May 2013. Heterogeneity in Computing Workshop, `http://www.netlib.org/utk/people/JackDongarra/SLIDES/hcw-0513.pdf`, accessed on 19th July 2017.

[61] J. Dongarra. Overview of High Performance Computing, November 2013. SC13, UTK Booth talk, Denver, Co, U.S.A., `http://www.netlib.org/utk/people/JackDongarra/SLIDES/sc13-UTK.pdf`, accessed on 19th July 2017.

[62] P. Dorozynski, P. Czarnul, A. Malinowski, K. Czurylo, L. Dorau, M. Maciejewski, and P. Skowron. Checkpointing of Parallel MPI Applications Using MPI One-sided API with Support for Byte-addressable Non-volatile RAM. *Procedia Computer Science*, 80:30 – 40, 2016. International Conference on Computational Science 2016, ICCS 2016, 6-8 June 2016, San Diego, California, USA.

[63] J. Duato, F. Igual, R. Mayo, A. Pea, E. Quintana-Ort, and F. Silla. An Efficient Implementation of GPU Virtualization in High Performance Clusters. In H.-X. Lin, M. Alexander, M. Forsell, A. Knpfer, R. Prodan, L. Sousa, and A. Streit, editors, *Euro-Par 2009 Parallel Processing Workshops*, volume 6043 of *Lecture Notes in Computer Science*, pages 385–394. Springer Berlin Heidelberg, 2010.

[64] J. Duato, A. Pena, F. Silla, R. Mayo, and E. Quintana-Orti. rCUDA: Reducing the Number of GPU-based Accelerators in High Performance Clusters. In *High Performance Computing and Simulation (HPCS), 2010 International Conference on*, pages 224–231, 28 2010-July 2.

[65] L. Durant, O. Giroux, M. Harris, and N. Stam. Inside Volta: The Worlds Most Advanced Data Center GPU, May 2017. `https://devblogs.nvidia.com/parallelforall/inside-volta/`, accessed on 19th July 2017.

[66] R. Farber. *Parallel Programming with OpenACC*. Morgan Kaufmann, Cambridge, MA, USA, 1st edition, November 2016. ISBN 978-0124103979.

[67] M. Flynn. *Encyclopedia of Parallel Computing*, chapter Flynn's Taxonomy, pages 689–697. Springer US, Boston, MA, 2011.

[68] S. S. Foley, D. Koepke, J. Ragatz, C. Brehm, J. Regina, and J. Hursey. OnRamp: A Web-portal for Teaching Parallel and Distributed Computing. *Journal of Parallel and Distributed Computing*, 105:138 – 149, 2017. Special issue: Keeping up with Technology: Teaching Parallel, Distributed and High-Performance Computing.

[69] I. Foster. Internet Computing and the Emerging Grid, December 2000. Nature Macmillan Publishers Ltd, nature web matters, `http://www.nature.com/nature/webmatters/grid/grid.html`, accessed on 19th July 2017.

[70] I. Foster, C. Kesselman, J. Nick, and S. Tuecke. The Physiology of the Grid: An Open Grid Services Architecture for Distributed Systems Integration. In *Open Grid Service Infrastructure WG*, June 22 2002. Global Grid Forum.

[71] I. Foster, C. Kesselman, and S. Tuecke. The Anatomy of the Grid: Enabling Scalable Virtual Organizations. *International Journal of High Performance Computing Applications*, 15(3):200–222, 2001.

[72] C. Funai, C. Tapparello, H. Ba, B. Karaoglu, and W. Heinzelman. Extending Volunteer Computing through Mobile Ad Hoc Networking. In *IEEE GLOBECOM Global Communications Conference Exhibition & Industry Forum*, Austin, TX, U.S.A., December 2014.

[73] A. Geist, A. Beguelin, J. Dongarra, W. Jiang, R. Manchek, and V. Sunderam. *PVM: Parallel Virtual Machine: A Users' Guide and Tutorial for Networked Parallel Computing*. MIT Press, Cambridge, MA, USA, 1994.

[74] S. Gogar. Intel Xeon Phi x200 Processor - Memory Modes and Cluster Modes: Configuration and Use Cases, December 2015. `https://software.intel.com/en-us/articles/intel-xeon-phi-x200-processor-memory-modes-and-cluster-modes-configuration-and-use-cases#MCDRAM-only_or_DDR-only_Mode`, accessed on 19th July 2017.

[75] R. W. Green. OpenMP* Thread Affinity Control, August 2012. Intel Developer Zone, `https://software.intel.com/en-us/articles/openmp-thread-affinity-control`, accessed on 19th July 2017.

[76] Gregg S. Process and Thread Affinity for Intel Xeon Phi Processors, April 2016. https://software.intel.com/en-us/articles/process-and-thread-affinity-for-intel-xeon-phi-processors-x200, accessed on 19th July 2017.

[77] W. Gropp, T. Hoefler, R. Thakur, and E. Lusk. *Using Advanced MPI: Modern Features of the Message-Passing Interface (Scientific and Engineering Computation)*. Scientific and Engineering Computation. The MIT Press, first edition, November 2014. ISBN 978-0262527637.

[78] W. Gropp, E. Lusk, and A. Skjellum. *Using MPI: Portable Parallel Programming with the Message-Passing Interface (Scientific and Engineering Computation)*. Scientific and Engineering Computation. The MIT Press, third edition, November 2014. ISBN 978-0262527392.

[79] S. Gupta, A. Agrawal, K. Gopalakrishnan, and P. Narayanan. Deep Learning with Limited Numerical Precision. In F. R. Bach and D. M. Blei, editors, *Proceedings of the 32nd International Conference on Machine Learning, ICML 2015, Lille, France, 6-11 July 2015*, volume 37 of *JMLR Workshop and Conference Proceedings*, pages 1737–1746. JMLR.org, 2015.

[80] GWT-TUD GmbH. Vampir - Performance Optimization. https://www.vampir.eu/, accessed on 19th July 2017.

[81] P. H. Hargrove and J. C. Duell. Berkeley Lab Checkpoint/Restart (BLCR) for Linux Clusters. *Journal of Physics: Conference Series*, 46(1):494, 2006.

[82] M. Harris. Optimizing Parallel Reduction in CUDA. NVIDIA Developer Technology. http://developer.download.nvidia.com/compute/cuda/1.1-Beta/x86_website/projects/reduction/doc/reduction.pdf, accessed on 19th July 2017.

[83] M. Harris. Unified Memory in CUDA 6, November 2013. https://devblogs.nvidia.com/parallelforall/unified-memory-in-cuda-6/, accessed on 19th July 2017.

[84] M. Harris. Using Shared Memory in CUDA C/C++, January 2013. https://devblogs.nvidia.com/parallelforall/using-shared-memory-cuda-cc/, accessed on 19th July 2017.

[85] M. Harris. GPU Pro Tip: CUDA 7 Streams Simplify Concurrency, January 2015. https://devblogs.nvidia.com/parallelforall/gpu-pro-tip-cuda-7-streams-simplify-concurrency/, accessed on 19th July 2017.

[86] C. He and P. Du. CUDA Performance Study on Hadoop MapReduce Clusters, 2010. University of Nebraska-Lincoln, CSE 930 Advanced Computer Architecture, http://www.slideshare.net/airbots/cuda-29330283.

[87] V. Holmes and I. Kureshi. Developing High Performance Computing Resources for Teaching Cluster and Grid Computing Courses. *Procedia Computer Science*, 51:1714 – 1723, 2015. International Conference On Computational Science, ICCS 2015.

[88] Indiana University. FAQ: Building CUDA-aware Open MPI, July 2016. https://www.open-mpi.org/faq/?category=buildcuda, accessed on 19th July 2017.

[89] Indiana University. FAQ: Running CUDA-aware Open MPI, July 2016. https://www.open-mpi.org/faq/?category=runcuda, accessed on 19th July 2017.

[90] Intel Corporation. Intel Cilk Plus. website, https://www.cilkplus.org/, accessed on 19th July 2017.

[91] Intel Corporation. Intel Threading Building Blocks. website, https://www.threadingbuildingblocks.org/, accessed on 19th July 2017.

[92] Intel Corporation. HeteroStreams 0.9 Programming Guide and API Reference. Intel Manycore Platform Software Stack 3.6, February 2016. https://01.org/sites/default/files/documentation/hstreams_reference_1.0.pdf, accessed on 19th July 2017.

[93] Intel Corporation. HStreams Tutorial. Hetero Streams Library 1.0, March 2016. https://01.org/sites/default/files/documentation/hstreams_1.0_tutorial.pdf, accessed on 19th July 2017.

[94] Intel Corporation. Intel MPI Library Developer Guide for Linux* OS, February 2017. https://software.intel.com/en-us/mpi-developer-guide-linux, accessed on 19th July 2017.

[95] R. Ioffe. OpenCL Tutorials, June 2015. Intel Developer Zone, https://software.intel.com/en-us/articles/opencl-tutorials, accessed on 19th July 2017.

[96] Ł. Jarzabek and P. Czarnul. Performance evaluation of unified memory and dynamic parallelism for selected parallel cuda applications. *The Journal of Supercomputing*, 73(12):5378–5401, Dec 2017.

[97] J. Jeffers and J. Reinders, editors. *High Performance Parallelism Pearls Volume Two: Multicore and Many-core Programming Approaches*. Morgan Kaufmann, Waltham, MA, USA, 1st edition, August 2015. ISBN 978-0128038192.

[98] S. Jones. Introduction to Dynamic Parallelism. NVIDIA Corporation, `http://on-demand.gputechconf.com/gtc/2012/presentations/S0338-GTC2012-CUDA-Programming-Model.pdf`, accessed on 19th July 2017.

[99] S. Kannan, A. Gavrilovska, K. Schwan, D. Milojicic, and V. Talwar. Using Active NVRAM for I/O Staging. In *Proceedings of the 2Nd International Workshop on Petascal Data Analytics: Challenges and Opportunities*, PDAC '11, pages 15–22, New York, NY, USA, 2011. ACM.

[100] K. Key and J. Ovall. A Parallel Goal-oriented Adaptive Finite Element Method for 2.5-D Electromagnetic Modelling. *Geophysical Journal International*, 186(1):137–154, 2011.

[101] D. B. Kirk and W. W. Hwu. *Programming Massively Parallel Processors.* Morgan Kaufmann, 3rd edition, December 2016. ISBN 978-0128119860.

[102] L. Howes (Editor). The OpenCL Specification Version: 2.1 Document Revision: 23, November 2015. Khronos OpenCL Working Group, `https://www.khronos.org/registry/cl/specs/opencl-2.1.pdf`, accessed on 19th July 2017.

[103] R. Landaverde, T. Zhang, A. K. Coskun, and M. Herbordt. An Investigation of Unified Memory Access Performance in CUDA. In *High Performance Extreme Computing Conference (HPEC), 2014 IEEE*, pages 1–6, Sept 2014.

[104] J. Larkin. 7 Powerful New Features in OpenACC 2.0, February 2014. `https://devblogs.nvidia.com/parallelforall/7-powerful-new-features-openacc-2-0/`, accessed on 19th July 2017.

[105] H.-F. Li, T.-Y. Liang, and J.-Y. Chiu. A Compound OpenMP/MPI Program Development Toolkit for Hybrid CPU/GPU Clusters. *The Journal of Supercomputing*, pages 1–25, 2013.

[106] M. Li and M. Baker. *The Grid - Core Technologies.* Wiley, Chichester, West Sussex, England, 2005.

[107] N. Losada, M. J. Martn, G. Rodrguez, and P. Gonzlez. Portable Application-level Checkpointing for Hybrid MPI-OpenMP Applications. *Procedia Computer Science*, 80:19 – 29, 2016. International Conference on Computational Science 2016, ICCS 2016, 6-8 June 2016, San Diego, California, USA.

[108] M. Lubin. Introduction into New Features of MPI-3.0 Standard. TCE DPD/SSG, Intel, `https://software.intel.com/sites/default/files/managed/ad/0f/Intel-MPI-library-new-features-and-performance-benchmarks-of-MPI3.0-standard.pdf`, accessed on 19th July 2017.

[109] J. Luitjens. CUDA Pro Tip: Always Set the Current Device to Avoid Multithreading Bugs, September 2014. https://devblogs.nvidia.com/parallelforall/cuda-pro-tip-always-set-current-device-avoid-multithreading-bugs/, accessed on 19th July 2017.

[110] D. Mackay. Optimization and Performance Tuning for Intel Coprocessors - Part 1: Optimization Essentials, November 2012. Intel Developer Zone, https://software.intel.com/en-us/articles/optimization-and-performance-tuning-for-intel-xeon-phi-coprocessors-part-1-optimization, accessed on 19th July 2017.

[111] T. MacWilliam and C. Cecka. CrowdCL: Web-based volunteer computing with WebCL. In *High Performance Extreme Computing Conference (HPEC), 2013 IEEE*, pages 1–6, Sept 2013.

[112] I. Massy, N. Pena, and M. Ney. Efficient Perfect Matched Layer for Hybrid MRTD-FDTD Computation with Lossy Dielectric Material Boundaries. *European physical journal - Applied physics*, 57(01), january 2012.

[113] P. Micikevicius. Multi-GPU Programming, May 2012. Developer Technology, NVIDIA, GPU Technology Conference, San Jose, California, USA.

[114] A. Munshi, B. Gaster, T. G. Mattson, J. Fung, and D. Ginsburg. *OpenCL Programming Guide.* Addison-Wesley Professional, Boston, MA, USA, 1st edition, 2011.

[115] B. Nichols, D. Buttlar, and J. P. Farrell. *Pthreads Programming. A POSIX Standard For Better Multiprocessing.* O'Reilly, Beijing, 1998.

[116] K. A. Nuaimi, N. Mohamed, M. A. Nuaimi, and J. Al-Jaroodi. A Survey of Load Balancing in Cloud Computing: Challenges and Algorithms. In *Proceedings of the 2012 Second Symposium on Network Cloud Computing and Applications*, NCCA '12, pages 137–142, Washington, DC, USA, 2012. IEEE Computer Society.

[117] A. Nukada, H. Takizawa, and S. Matsuoka. NVCR: A Transparent Checkpoint-Restart Library for NVIDIA CUDA. In *2011 IEEE International Symposium on Parallel and Distributed Processing Workshops and Phd Forum*, pages 104–113, May 2011.

[118] NVIDIA. Whitepaper. NVIDIA Tesla P100. The Most Advanced Datacenter Accelerator Ever Built Featuring Pascal GP100, the Worlds Fastest GPU. https://images.nvidia.com/content/pdf/tesla/whitepaper/pascal-architecture-whitepaper.pdf, accessed on 19th July 2017.

[119] NVIDIA. Multi-Process Service, May 2015. vR352, https://docs.nvidia.com/deploy/pdf/CUDA_Multi_Process_Service_Overview.pdf, accessed on 19th July 2017.

[120] NVIDIA. NVIDIA GPUDirect, 2016. https://developer.nvidia.com/gpudirect, accessed on 19th July 2017.

[121] NVIDIA. OpenACC Toolkit, 2016. https://developer.nvidia.com/openacc-toolkit, accessed on 19th July 2017.

[122] NVIDIA. CUDA C Programming Guide, September 2017. version 9.0, http://docs.nvidia.com/cuda/pdf/CUDA_C_Programming_Guide.pdf, accessed on 5th November 2017.

[123] NVIDIA. CUDA Runtime API. API Reference Manual, July 2017. https://docs.nvidia.com/cuda/pdf/CUDA_Runtime_API.pdf, accessed on 5th November 2017.

[124] OpenACC-Standard.org. OpenACC Programming and Best Practices Guide, June 2015. https://www.openacc.org/sites/default/files/inline-files/OpenACC_Programming_Guide_0.pdf.

[125] OpenACC-Standard.org. The OpenACC Application Programming Interface. Version 2.5, October 2015. http://www.openacc.org/sites/default/files/OpenACC_2pt5.pdf, accessed on 19th July 2017.

[126] OpenMP Architecture Review Board. OpenMP Application Programming Interface. Examples, November 2013. Version 4.0.0, http://openmp.org/mp-documents/OpenMP4.0.0.Examples.pdf, accessed on 19th July 2017.

[127] OpenMP Architecture Review Board. OpenMP Application Programming Interface, November 2015. Version 4.5, http://www.openmp.org/mp-documents/openmp-4.5.pdf, accessed on 19th July 2017.

[128] Oracle. Multithreaded Programming Guide, 2010. https://docs.oracle.com/cd/E19455-01/806-5257/index.html, accessed on 19th July 2017.

[129] P. Pacheco. *An Introduction to Parallel Programming.* Morgan Kaufmann Publishers Inc., San Francisco, CA, USA, 1st edition, 2011.

[130] V. Pankratius, A.-R. Adl-Tabatabai, and W. Tichy, editors. *Fundamentals of Multicore Software Development.* Chapman & Hall/CRC Computational Science. CRC Press, Boca Raton, FL, USA, 1st edition, July 2017. ISBN 978-1138114371.

[131] PassMark Software. CPU Benchmarks. http://cpubenchmark.net/, accessed on 19th July 2017.

[132] D. K. Patel, D. Tripathy, and C. Tripathy. Survey of Load Balancing Techniques for Grid. *Journal of Network and Computer Applications*, 65:103 – 119, 2016.

[133] J. Proficz and P. Czarnul. *Performance and Power-Aware Modeling of MPI Applications for Cluster Computing*, pages 199–209. Springer International Publishing, Cham, Switzerland, 2016.

[134] R. Rahman. *Intel Xeon Phi Coprocessor Architecture and Tools. The Guide for Application Developers.* Apress, New York, NY, USA, September 2013. ISBN 978-1-4302-5926-8.

[135] J. Reinders. An Overview of Programming for Intel Xeon Processors and Intel Xeon Phi Coprocessors, November 2012. Intel Developer Zone, `https://software.intel.com/en-us/articles/an-overview-of-programming-for-intel-xeon-processors-and-intel-xeon-phi-coprocessors`, accessed on 19th July 2017.

[136] J. Reinders and J. Jeffers, editors. *High Performance Parallelism Pearls Volume One: Multicore and Many-core Programming Approaches.* Morgan Kaufmann, Waltham, MA, USA, 1st edition, November 2014. ISBN 978-0128021187.

[137] S. Rennich. CUDA C/C++. Streams and Concurrency, 2011. NVIDIA, `http://on-demand.gputechconf.com/gtc-express/2011/presentations/StreamsAndConcurrencyWebinar.pdf`, accessed on 19th July 2017.

[138] R. Reyes, I. Lopez, J. J. Fumero, J. L. Grillo, and F. de Sande. accULL. The Open Source OpenACC Implementation. `https://bitbucket.org/ruyman/accull/downloads/`, accessed on 19th July 2017.

[139] R. Reyes, I. Lopez-Rodriguez, J. J. Fumero, and F. de Sande. *accULL: An OpenACC Implementation with CUDA and OpenCL Support*, pages 871–882. Springer Berlin Heidelberg, Berlin, Heidelberg, 2012.

[140] G. Rodriguez, M. J. Martin, P. Gonzalez, J. Tourino, and R. Doallo. CPPC: a Compiler-assisted Tool for Portable Checkpointing of Message-passing Applications. *Concurrency and Computation: Practice and Experience*, 22(6):749–766, 2010.

[141] P. Rosciszewski, P. Czarnul, R. Lewandowski, and M. Schally-Kacprzak. KernelHive: a New Workflow-based Framework for Multilevel High Performance Computing using Clusters and Workstations with CPUs and GPUs. *Concurrency and Computation: Practice and Experience*, 28(9):2586–2607, 2016. cpe.3719.

[142] F. Roth. Quick Start Guide for the Intel Xeon Phi Processor x200 Product Family, April 2016. `https://software.intel.com/en-us/articles/quick-start-guide-for-the-intel-xeon-phi-processor-x200-product-family`, accessed on 19th July 2017.

[143] N. Sakharnykh. The Future of Unified Memory, April 2016. GPU Technology Conference, `http://on-demand.gputechconf.com/gtc/2016/presentation/s6216-nikolay-sakharnykh-future-unified-memory.pdf`, accessed on 19th July 2017.

[144] M.-R. Sancho. BSC Best Practices in Professional Training and Teaching for the HPC Ecosystem. *Journal of Computational Science*, 14:74 – 77, 2016. Special issue: The Route to Exascale: Novel Mathematical Methods, Scalable Algorithms and Computational Science Skills.

[145] J. Sanders and E. Kandrot. *CUDA by Example: An Introduction to General-Purpose GPU Programming.* Addison-Wesley Professional, Boston, MA, USA, 2010. ISBN-13: 978-0131387683.

[146] L. P. Santos, V. Castro, and A. Proença. *Evaluation of the Communication Performance on a Parallel Processing System*, pages 41–48. Springer Berlin Heidelberg, Berlin, Heidelberg, 1997.

[147] C. Sarris and L. Katehi. An Efficient Numerical Interface between FDTD and Haar MRTD-formulation and Applications. *Microwave Theory and Techniques, IEEE Transactions on*, 51(4):1146–1156, Apr 2003.

[148] C. Sarris, K. Tomko, P. Czarnul, S. Hung, R. Robertson, D. Chun, E. Davidson, and L. Katehi. Multiresolution Time Domain Modeling for Large Scale Wireless Communication Problems. In *Proceedings of the 2001 IEEE AP-S International Symposium on Antennas and Propagation*, volume 3, pages 557–560, 2001.

[149] J. A. Shamsi, N. M. Durrani, and N. Kafi. Novelties in Teaching High Performance Computing. In *2015 IEEE International Parallel and Distributed Processing Symposium Workshop*, pages 772–778, May 2015.

[150] S. S. Shende and A. D. Malony. The Tau Parallel Performance System. *Int. J. High Perform. Comput. Appl.*, 20(2):287–311, May 2006.

[151] L. Shi, H. Chen, and T. Li. *Hybrid CPU/GPU Checkpoint for GPU-Based Heterogeneous Systems*, pages 470–481. Springer Berlin Heidelberg, Berlin, Heidelberg, 2014.

[152] V. Silva. *Grid Computing for Developers.* Charles River Media, Hingham, Mass, 2005.

[153] H. Takizawa, K. Koyama, K. Sato, K. Komatsu, and H. Kobayashi. CheCL: Transparent Checkpointing and Process Migration of OpenCL

Applications. In *2011 IEEE International Parallel Distributed Processing Symposium*, pages 864–876, May 2011.

[154] H. Takizawa, K. Sato, K. Komatsu, and H. Kobayashi. CheCUDA: A Checkpoint/Restart Tool for CUDA Applications. In *2009 International Conference on Parallel and Distributed Computing, Applications and Technologies*, pages 408–413, Dec 2009.

[155] The IEEE and The Open Group. The Open Group Base Specifications Issue 7, IEEE Std 1003.1-2008, 2016 Edition, 2016. `http://pubs.opengroup.org/onlinepubs/9699919799/`, accessed on 19th July 2017.

[156] J. Tourino, M. J. Martin, J. Tarrio, and M. Arenaz. A Grid Portal for an Undergraduate Parallel Programming Course. *IEEE Transactions on Education*, 48(3):391–399, Aug 2005.

[157] A. Tousimojarad and W. Vanderbauwhede. Comparison of Three Popular Parallel Programming Models on the Intel Xeon Phi. In *Euro-Par 2014: Parallel Processing*. Springer.

[158] R. Tsuchiyama, T. Nakamura, T. Iizuka, A. Asahara, J. Son, and S. Miki. *The OpenCL Programming Book*. Fixstars Corporation and Impress Japan Corporation, 2012. `https://www.fixstars.com/en/opencl/book/`.

[159] UNICORE Team. UNICORE Rich Client User Manual, October 2016. version 7.3.2, `https://www.unicore.eu/docstore/urc-7.3.2/manual.html`, accessed on 19th July 2017.

[160] University of California. BOINC. `http://boinc.berkeley.edu/`, accessed on 3rd July 2017.

[161] University of Chicago. Globus Toolkit Homepage. `http://toolkit.globus.org/toolkit/`, accessed on 19th July 2017.

[162] University of Chicago. GT 6.0 GRAM5: User's Guide. `http://toolkit.globus.org/toolkit/docs/latest-stable/gram5/user/index.html`, accessed on 19th July 2017.

[163] University of Chicago. GT 6.0 GridFTP: User's Guide. `http://toolkit.globus.org/toolkit/docs/latest-stable/gridftp/user/index.html`, accessed on 19th July 2017.

[164] University of Chicago. Security: GSI C User's Guide. `http://toolkit.globus.org/toolkit/docs/latest-stable/gsic/user/index.html`, accessed on 19th July 2017.

[165] R. van der Pas. OpenMP Tasking Explained, November 2013. SC13 Talk at OpenMP Booth, Santa Clara, CA, USA, `http://openmp.org/wp-content/uploads/sc13.tasking.ruud.pdf`, accessed on 19th July 2017.

[166] A. Vladimirov, R. Asai, and V. Karpusenko. *Parallel Programming and Optimization with Intel Xeon Phi Coprocessors.* Colfax International, Sunnyvale, CA, USA, May 2015. ISBN 978-0-9885234-0-1.

[167] J. P. Walters and V. Chaudhary. *Application-Level Checkpointing Techniques for Parallel Programs*, pages 221–234. Springer Berlin Heidelberg, Berlin, Heidelberg, 2006.

[168] W. Wang, W. Liu, and Z. Wang. *Survey of Load Balancing Strategies on Heterogeneous Parallel System*, pages 583–589. Springer Netherlands, Dordrecht, 2011.

[169] X. Wang, Y. Hao, and C.-H. Chu. Stability Comparison between Multi-Resolution Time-Domain (MRTD) and Finite-Difference Time-Domain (FDTD) Techniques. In *2008 Loughborough Antennas and Propagation Conference*, pages 269–272, March 2008.

[170] B. Wilkinson and M. Allen. *Parallel Programming: Techniques and Applications Using Networked Workstations and Parallel Computers.* Pearson, Upper Sadle River, NJ, USA, 2nd edition edition, March 2004. ISBN 978-0131405639.

[171] M. H. Willebeek-LeMair and A. P. Reeves. Strategies for Dynamic Load Balancing on Highly Parallel Computers. *IEEE Trans. Parallel Distrib. Syst.*, 4(9):979–993, Sept. 1993.

Index